PROBABILITY APPLICATIONS IN MECHANICAL DESIGN

MECHANICAL ENGINEERING
A Series of Textbooks and Reference Books

Founding Editor

L. L. Faulkner

*Columbus Division, Battelle Memorial Institute
and Department of Mechanical Engineering
The Ohio State University
Columbus, Ohio*

Additional Volumes in Preparation

Mechanical Engineering Software

PROBABILITY APPLICATIONS IN MECHANICAL DESIGN

FRANKLIN E. FISHER

JOY R. FISHER

Loyola Marymount University
Los Angeles, California

MARCEL DEKKER, INC.　　　　　　　　　　NEW YORK · BASEL

ISBN: 0-8247-0260-3

This book is printed on acid-free paper.

Headquarters
Marcel Dekker, Inc.
270 Madison Avenue, New York, NY 10016
tel: 212-696-9000; fax: 212-685-4540

Eastern Hemisphere Distribution
Marcel Dekker AG
Hutgasse 4, Postfach 812, CH-4001, Basel, Switzerland
tel: 41-61-261-8482; fax: 41-61-261-8896

World Wide Web
http://www.dekker.com

The publisher offers discounts on this book when ordered in bulk quantities. For more information, write to Special Sales/Professional Marketing at the headquarters address above.

Current printing (last digit):
10 9 8 7 6 5 4 3 2 1

PRINTED IN THE UNITED STATES OF AMERICA

Preface

This book is intended for use by practicing engineers in industry, but formatted with examples and problems for use in a one-semester graduate course.

Chapter 1 provides the data reduction techniques for fitting experimental failure data to a statistical distribution. For the purposes of this book only normal (Gaussian) and Weibull distributions are considered, but the techniques can be expanded to include other distributions, including non-parametric distributions.

The main part of the book is Chapter 2, which applies probability and computer analysis to fatigue, design, and variations of both. The essence of this chapter is the ideas presented in *Metal Fatigue* (1959) edited by George Sines and J. L. Waisman and considers the problem of having to deal with a limited amount of engineering data. The discussions of fatigue by Robert C. Juvinall in *Stress, Strain, Strength* (1967) and by J. H. Faupel and F. E. Fisher in *Engineering Design* (1981), as well as the books by Edward Haugen (1968) on the variation of parameters in fatigue, are successfully combined into a single treatment of fatigue. This book is an extension of Haugen's book *Probabilistic Mechanical Design* (1980) with applications.

The concepts of optimization are developed in Chapter 3. The technique of geometric programming is presented and solutions to sample problems are compared with computer-generated non-linear programming solutions. Reliability, the topic Chapter 4, is developed for mechanical systems and some failure rate data is presented as it can be hard to find.

The book is influenced by the consulting work I performed at Hughes Aircraft Co. from 1977 to 1993. Some of the examples are drawn from this effort.

Joy Fisher, worked in computer programming in the 1980s and 1990s keeping track of the changing state of the art in computing and writing for sections in this book dealing with programming.

This book was roughed out on a sabbatical leave in 1994 from class notes and in a summer institute taught by Edward Haugen in the early 1970s. Credit also goes to many students from industry who labored to understand and use the information.

The editorial and secretarial assistance of Ms. Cathy Herrera is gratefully acknowledged.

Franklin E. Fisher
Joy R. Fisher

Contents

List of Symbols

B	A test sample Weibull β from a plot or computer
$C\left(\dfrac{n}{r}\right)$	Combinations
$C - c(x_1 \ldots x_n)$	Criterion function
$C_v - \dfrac{z}{\mu} \times 100$	A percentage coefficient of variation
$F_m - f_m(x_1 \ldots x_n)$	Functional constraints
$F(x) - 1 - \displaystyle\int_{-\infty}^{x} f(x)dx$	Gaussian failure
$f(A)$	Resisting capacity
$f(a)$	Applied load
$f(t)$	Failures with respect to time
$f(x)$	Test data fitted to a Gaussian curve
$f_i(x_i)$	Gaussian curve values for the middle of each cell width
$G(x) - 1 - \displaystyle\int_{\gamma}^{x} g(x)dx$	Weibull failure
$g(x)$	Test data fitted to a Weibull curve
$g_i(x_i)$	Weibull curve values for middle of each cell width
(K)	Number of cells Sturges Rule
K_f	K_t Corrected for material
K_F	Severity factors
K_L	Life-expectancy severity factor
$K(n)$	Bounds on the Weibull Line
K_t	Theoretical stress concentration factors
k_a	Surface condition
k_b	Size and shape
k_c	Reliability

k_d	Temperature
k_e	Stress concentration
k_f	Residual stress
k_g	Internal structure
k_h	Environment
k_i	Surface treatment and hardening
k_j	Fretting
k_k	Shock or vibration loading
k_l	Radiation
k_m	Corrections above 10^7 cycles
$MTTF - \dfrac{1}{\lambda}$	Constant failure rate $e^{-\lambda t}$
N	The total number of test samples
$N - A/B$	Safety factor for $\sigma_r - \sigma_m$ curve
$N_f(t)$	Items failed in service by time t
$N_s(t)$	Items in service at time t
$P(A)$	Probability of A occurring
$P(\bar{A})$	Probability of A not occurring
$P(A + B)$	Probability A or B can happen or both
$P(AB)$	Probability A happens followed by B
$P(A/B)-\text{``}B\text{''}$	Happened the probability that it was followed by "A"
$P(B/A)-\text{``}A\text{''}$	Happened, the probability it was followed by "B"
P_f	Percent failures
$Q(t) - N_f(t)/N$	Probability failure items failed versus total
q	Notch sensitivity factor
$R - \dfrac{\sigma_{\min}}{\sigma_{\max}}$	Stress ratio Chapter 2
(R)	Range of data Sturges Rule Chapter 1
$R(t) - \dfrac{N_s(t)}{N}$	Reliability (items in service versus total)
s	Gaussian standard deviation calculated from test samples
t	Students distribution (Appendix E)
$t - \dfrac{\mu_A - \mu_a}{[(\breve{z}_A)^2 + (\breve{z}_a)^2]^{1/2}}$	coupling equation
t_G	Generic Life-Expectancy Distribution
$(W) - R/K$	Cell width Sturges Rule
$X\breve{z}_c$	Standard deviations in a card sort
$\bar{x}$	A sample Gaussian mean calculated from test samples
Y	Cold working improvements for k_f

Y	A scaling factor for Weibull plotting
$\breve{z}$	Is the Gaussian standard deviation for an infinite sample size
$\breve{Z}\psi$	Standard deviation for a function $\psi\,(x, y, z, \ldots)$
α	One sided tolerance limit
β	Is a Weibull shape parameter for infinite sample size
x^2	Chi-Square Distribution
χ	A scaling factor for Gaussian plotting
Δ	A test sample Weibull δ from a plot or computer
δ	Is a Weibull scale parameter for infinite sample size
$\Delta\varepsilon$	Strain low cycle fatigue
λ	Is a Weibull locations parameter less than the lowest value of the infinite data
$\lambda - \dfrac{\Delta N_f}{N_s}\dfrac{1}{\Delta t}$	Failure rate (failures per hour)
$\lambda - \lambda_G K_F$	(Appendix D)
λ_G	Generic fail-rate distributions
λ_m	Lagrangian multiplier
μ	Is the Gaussian mean for an infinite sample size
Θ^β	Another form of the Weibull δ
θ	A test sample Weibull Θ from a plot or computer
σ_e	Corrected specimen endurance
$\sigma_m - \dfrac{\sigma_{\max} + \sigma_{\min}}{2}$	Mean stress
σ'_m	$\sqrt{\sigma_{xm}^2 - \sigma_{xm}\sigma_{ym} + \sigma_{ym}^2 + 3\tau_{xym}^2}$
$\sigma_r - \dfrac{\sigma_{\max} - \sigma_{\min}}{2}$	Reversal or amplitude stress
σ'_r	$\sqrt{\sigma_{xr}^2 - \sigma_{xr}\sigma_{yr} + \sigma_{yr}^2 + 3\tau_{xyr}^2}$
σ_y	Yield strength
$(\mu_1, \breve{z}_1)$	Mean, standard deviation for a variable

1
Data Reduction

Data for load carrying material properties can be modelled using any probability distribution function. Statistical goodness-of-fit tests should be applied to determine if the data set could be randomly drawn from that distribution. Modelling has progressed beyond a simple two parameter $(\mu, \check{z})$ Gaussian distribution. This book treats the three parameter (δ, β, γ) Weibull distribution, as well as the traditional Gaussian distribution.

Many authors relegate the subject of data reduction to an appendix at the back of the book. In the opinion of the authors, the topic deserves much more attention.

I. REDUCTION OF RAW TABULATED TEST DATA OR PUBLISHED BAR CHARTS

A computer program such as SAS (Statistical Analysis System) statistical software or other compatible software is used to fit test data to a Gaussian curve.

$$f(x) = \frac{1}{\check{z}\sqrt{2\pi}} \exp\left[-\frac{1}{2}\left(\frac{x-\mu}{\check{z}}\right)^2\right] \tag{1.1}$$

where $-\infty \leq x \leq +\infty$ with

μ–is the mean for an infinite sample size
$\check{z}$–is the standard deviation for an infinite sample size.
The program also fits data to a Weibull curve,

$$g(x) = \beta\frac{[x-\gamma]^{\beta-1}}{\delta} \exp\left[-\frac{[x-\gamma]^\beta}{\delta}\right] \tag{1.2}$$

where $\gamma \leq x \leq +\infty$ and g (x) $=$ 0 when x $\leq \gamma$ and

δ–is a scale parameter for infinite sample size
β–is a shape parameter for infinite sample size
γ–is a threshold parameter

The computer solves for the Weibull parameters as well as the Gaussian mean and standard deviation for a set of individual values from mechanical testing or published bar charts with more than one or two samples at a given value for the mid point of the bar (cell width).

Some computer software solves for only two Weibull parameters δ and β while γ is set to zero. The failure curves $f(x)$ and $g(x)$ are used to generate the reliability

$$F(x) = 1 - \int_{-\infty}^{x} f(x)dx \tag{1.3}$$

$$G(x) = 1 - \int_{\gamma}^{x} g(x)dx \tag{1.4}$$

The Weibull $G(x)$, $g(x)$ and Gaussian $F(x)$ $f(x)$ are unity curves with values from zero to one. The Weibull and Gaussian curves are used throughout this book except in chapter four where the constant failure-rate for reliability is introduced to explain the wear and tear on machinery. The computer calculations for Eqs. (1.1) and (1.2) allows the individual data to be sorted or listed as a bar chart. Sturges Rule [1.12,1.16] presents an acceptable means of plotting data on linear coordinates, where, the data is grouped in cells of width w, over the range R, of the data.

1.   Number of cells $(K) = 1 + 3.3 \log_{10} N$ where N is the number of individual data points. Grouped data from bar charts are already partitioned as presented in the data source, so the steps outlined for using Sturges Rule will not apply: however, the number of cells can be checked.
2.   Range (R) = maximum value minus the minimum value.
3.   Cell width $(W) = R/K$.

The number of cells can be rounded off, say -7.2 is 7 cells and -7.8 is 8 cells. Then using a sorted list of values partition the data into the number of cells and plot N_i for each cell versus the value of the data in the center of a cell width.

The test sample Gaussian curve values are calculated for the middle of each cell width x_i so from Eq. (1.1).

$$f_i(x_i) = \frac{1}{\check{z}\sqrt{2\pi}} \exp\left[-\frac{1}{2}\left(\frac{x_i - \mu}{\check{z}}\right)^2\right] \tag{1.5}$$

If there are 8 cells the total is scaled up to reflect the total number of test data.

$$N = \chi \sum_{i=1}^{i=8} f_i(x_i) \tag{1.6}$$

or

$$\chi = \frac{N}{\Sigma f_i(x_i)} \tag{1.7}$$

Equation (1.7) is a scaling factor which at each midpoint expands the Gaussian unity curve-value $f_i(x_i)$ to reflect the actual data so that

$$\Sigma N_i = N \tag{1.8}$$

where N is the total number of test samples and is a check on the calculations.

The Weibull equation is also expanded and plotted on the same bar chart with the Gaussian curve.

Equation (1.2) for the midpoints of the cells i is expressed as

$$g_i(x_i) = \frac{\beta[x_i - \gamma]^{\beta-1}}{\delta} \exp\left[\frac{[x_i - \gamma]^{\beta}}{\delta}\right] \tag{1.9}$$

Then

$$N = Y \Sigma g_i(x_i) \tag{1.10}$$

Solving for Y by summing on i

$$Y = \frac{N}{\Sigma g_i(x_i)} \tag{1.11}$$

which expands each midpoint of the Weibull plot so that

$$N = \Sigma N_i \tag{1.12}$$

Again a check on calculations is made.

The data may be further checked by plotting the sum of the failures

$$Y \int_0^x g(x)dx = x \int_0^x f(x)dx = \frac{1}{N}\Sigma N_i \tag{1.13}$$

Verus a log scale on the right hand side. This is done by adding from $i=1$ to the desired cell, dividing by N and plotting this value at the end of cell i. This is a probability of failure and when it is 0.5 it gives a good check on the Gaussian mean and the peak of the Weibull curve. The value 0.5 is also called a 50th percentile for the data and shows how the data is skewed.

II. WEIBULL EQUATION VARIATIONS

Equation (1.2) may also be stated for infinite sample size

$$g(x) = \frac{\beta}{\Theta}\left[\frac{x-\gamma}{\Theta}\right]^{\beta-1}\exp\left[-\left(\frac{x-\gamma}{\Theta}\right)^{\beta}\right] \tag{1.14}$$

Comparing Eqs. (1.2) and (1.14)

$$\Theta^{\beta} = \delta \tag{1.15}$$

Published Weibull parameters have β which is the same for Eqs. (1.2) and (1.14) but, will have either Θ or δ; and may have a value for γ or analyzed as if it is zero. Also Eq. (1.4) will be

$$G(x) = 1 - \exp\left[-\frac{(x-\gamma)^{\beta}}{\delta}\right] = 1 - \exp\left[-\left(\frac{(x-\gamma)}{\Theta}\right)^{\beta}\right] \tag{1.16}$$

The distribution is called:

(a) Two parameter Weibull when β and Θ or δ given and γ equal zero and computer calculated.
(b) Three parameter Weibull the same β and Θ or δ and γ is the lowest data value or selected by the computer.

III. PLOTTING RAW TABULATED TEST DATA OR USING PUBLISHED BAR CHARTS

A. Weibull

The value of γ can be determined from the finite sample data and is smaller than the minimum value.

The finite sample plotting is performed on Weibull paper which is a ln ln versus ln graph paper as discussed in Appendix A. The data is divided into cells using Sturges Rule (Section 1.1).

Percent failures

$$p_f = \frac{\Sigma N_i}{N} \times 100 \tag{1.17}$$

are plotted at the end of each cell and a best fit straight line is drawn through the data and β estimated. The Weibull form for the line is

$$p_f = 1 - \exp\left[-\frac{(x-\gamma)^{\beta}}{\delta}\right] = 1 - \exp\left[-\left(\frac{x-\gamma}{\Theta}\right)^{\beta}\right] \tag{1.18}$$

note that

$$(x - \gamma) = \delta^{1/\beta} = \Theta \tag{1.19}$$

when

$$p_f = 1 - \exp[-1] = 0.632 \tag{1.20}$$

then drawing a line at $p_f = 0.632$ intersecting the best fit straight line the horizontal axis is $(x-\gamma)$. Since γ and β are known the values for δ and Θ can be calculated. The operations are developed more thoroughly in [1.1–1.3]. Abernathy [1.1,1.2] especially has published two books with many plotted engineering examples.

Now an indication of what kind of distributions the βs may indicate [1.16].

1. $\beta = 1$ Exponential distribution with constant rate failure.
2. $\beta = 2$ Rayleigh distribution.
3. $\beta = 3$ Log-normal distribution with normal wear out.
4. $\beta \approx 5$ or more normal distribution or Gaussian.

B. Gaussian

The data is divided into cells using Sturges Rule and plotted on normal probability paper and is plotted as a percent p_f. The peak at 50% should be the mean. The values of 6.3 and 93.3 are plus and minus three standard deviations about the mean. These values may be calculated from the plotted data.

The same $f(x)$ and $g(x)$ curves described in Section 1.1 can be calculated. If only estimates are required the plotted data may suffice but often more information is requested.

IV. CONFIDENCE LEVELS

The values of δ, β, γ and $\check{z}$ and μ are calculated or derived from plots for a given sample size N. The test sample size, $N \leq 30$, is normally called a small sample.

Chapter 1

A. Gaussian Distribution

1. Students t Distribution

A sample mean $\bar{x}$ is calculated from test samples with standard deviation of s. It is desired to find the infinite sample mean μ to a confidence of 95% (higher levels may also be used).

$$\bar{x} - t_{0.975} \frac{s}{N-1} \leq \mu \leq \bar{x} + t_{0.975} \frac{s}{N-1} \tag{1.21}$$

Note that 2.5% is in each tail to make 95% therefore for each side $t_{0.975}$ is used. The degrees of freedom (d.f) is $N-1$ and the values of t for 2.5% can be read in Table E.2.

2. Chi-Square Distribution

$$x^2 = \frac{Ns^2}{\sigma^2} \tag{1.22}$$

where $N-1$ is the degrees of freedom d.f. calculated from the test sample of N. σ^2 is the infinite sample size standard deviation. Here again 95% confidence $x_{0.975}$ and $x_{0.025}$ are used so that

$$\frac{s\sqrt{N}}{x_{0.025}} \leq \sigma \leq \frac{s\sqrt{N}}{x_{0.975}} \tag{1.23}$$

values for $x_{0.975}^2$ and $x_{0.025}^2$ are read from Table E.3. Here σ is used for $\check{z}$ in Eq. (1.1)

3. One Sided Tolerance Limit

$$\sigma = \bar{x} - Ks \tag{1.24}$$

$\bar{x}$, s are from limited sample. Fig. E.1 allows selection of K when the sample size and percent confidence is known. For example, choosing σ so that 90% of the experimental values are greater than σ with a 95% confidence limit.

4. Estimate of the Mean

The estimate of the mean, μ, and standard deviation σ or $\check{z}$ are discussed by Dixon and Massey [1.8], where small sample size values are arranged in ascending order $x_1, x_2, \ldots, x_n$ and number $n < 20$. The estimate for the μ and σ or $\check{z}$ are listed in Tables E.4–E.6. The values do not have a percent confidence attached but are for the infinite sample size.

The following example shows the good and bad features of a small sample size and is presented to show the variation in some calculations.

EXAMPLE 1.1. In order to illustrate the concepts for estimating μ and σ or $\check{z}$ for small samples, $N < 30$, select two test stress values 40,000 psi and 45,000 psi. These are from Eq. E.1 in the range of ultimates for 6061-T6 aluminum, therefore, the final answers can be compared to MILHDBK 5F [1.18] for 0.010–0.249 sheet

$$\text{A Basis} - \sigma_{ut} = 42 \text{ Ksi } \sigma_{yt} = 36 \text{ Ksi} \tag{1.25}$$

$$\text{B Basis} - \sigma_{ut} = 43 \text{ Ksi } \sigma_{yt} = 38 \text{ Ksi} \tag{1.26}$$

The mean $\bar{x}$ and standard deviation, s, for test samples of two will be calculated

$$\text{w-range} = 45{,}000 - 40{,}000 = 5000 \text{ psi} \tag{1.27}$$

$$\bar{x}\text{-means} = \frac{\Sigma x_i}{N} = \frac{1}{2}(45{,}000 + 40{,}000) = 42{,}500 \text{ psi} \tag{1.28}$$

$$\text{s-STD Deviation} = \left[\frac{\Sigma(x_i - \bar{x})^2}{N-1}\right]^{1/2} =$$

$$\left[\frac{(45 \times 10^3 - 42.5 \times 10^3)^2 + (40 \times 10^3 - 42.5 \times 10^3)^2}{2-1}\right]^{1/2} \tag{1.29}$$

$$= 3{,}536 \text{ psi (this for N = 1 will not work out)}$$

Another estimate, range $= 6s$, and s is 833 psi. This value will be used and checked against the final σ or $\check{z}$ from [1.8]. Now to find μ from Eq. (1.21) for 95% confidence

$$\bar{x} - t_{0.975}\frac{s}{N-1} \leq \mu \leq \bar{x} + t_{0.975}\frac{s}{N-1} \tag{1.30}$$

$$\text{d.f.} = 2 - 1 = 1$$

$$t_{0.975} = t_{0.025} = 12.706 \text{ (Table E.2.)}$$

$$42{,}500 - 12.706\frac{833}{1} \leq \mu \leq 42{,}500 + 12.706\frac{833}{1}$$

$$31{,}916 \text{ psi} \leq \mu \leq 53{,}084 \text{psi}$$

range infinite sample size for 95% confidence from a sample size of two. Next estimate σ or $\check{z}$ from the data using Eq. (1.23) for 95% confidence

$$\frac{s\sqrt{N}}{x_{0.025}} \leq \sigma \leq \frac{s\sqrt{N}}{x_{0.975}} \tag{1.31}$$

from Table E.3. for d.f. $= N-1 = 2-1 = 1$

$x_{0.025}^2 = 0.0009$	$x_{0.025} = 0.0313$
$x_{0.975}^2 = 5.02$	$x_{0.975} = 2.2405$

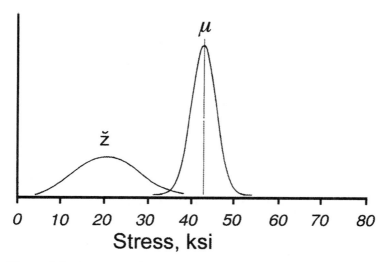

Figure 1.1 Mean, μ and standard deviation σ or $\check{z}$ for 95% confidence and infinite sample size.

substituting

$$\frac{833\sqrt{2}}{2.2405} \leq \sigma \leq \frac{833\sqrt{2}}{0.0313}$$

525.8 psi $\leq \sigma \leq$ 37,631 psi

Now sketch the values in (Fig. 1.1). As it will be seen, some features will present contradictions

Note from Fig. 1.1

$\mu - 3s =$ negative values for most of the final results.

If a design allowable is picked for $N = 2$ the approach in Eq. (1.24) and Fig. 1.1 for 95% confidence

$$\alpha = \bar{x} - Ks \tag{1.32}$$

A–basis 99% of values greater than α_A with K = 37.094 (Table E.1)
B–basis 90% of values greater than α_B with K = 20.581 (Table E.1)
$\alpha_A = 42,500 - 37.094 \ (833) = 11,600$ psi for $\bar{x}$ values
$\alpha_B = 42,500 - 20.581 \ (833) = 25,356$ psi for $\bar{x}$ values

This is better but still not great. An approach [1.8] will be attempted using Table E.6. Average of best two

$$\mu = 1/2(x_1 + x_2) = 1/2(40,000 + 45,000) = 42.500 \tag{1.33}$$

from Table E.4

$$N = 2$$
$$0.571\sigma^2 = 0.886 \, w$$
$$\text{Range w} = 45,000 - 40,000 = 5,000 \, \text{psi}$$
$$\sigma = \sqrt{\frac{0.886(5,000)}{0.571}}$$
$$\sigma = 88.081 \, \text{psi}$$

(1.34)

The calculation Table E.5

$$\sigma = \sqrt{\frac{0.8862(5,000)}{0.571}}$$
$$\sigma = 88.091 \, \text{psi}$$

If A and B values for two samples are calculated with $K_A = 37.094$, from Table E.1 and $K_B = 20.581$.

$$\alpha_A = \mu - K_A \, \sigma$$
$$\alpha_A = 42,500 - 37.094 \, (88.09 \, \text{psi})$$
$$\alpha_A = 39,232 \, \text{psi}$$

(1.35)

$$\alpha_B = \mu - K_B \, \sigma$$
$$\alpha_B = 42,500 - 20.581 \, (88.09 \, \text{psi})$$
$$\alpha_B = 40,687 \, \text{psi}$$

(1.36)

These are compared to actual values of 42,000 and 43,000 for A and B basis. These are now closer and could be used for designing.

5. Larger Data Samples $N > 30$

When more data is available, say >20–30, the estimates in Example 1.1 get better for the t and x calculations to get a mean and standard deviation. When $\bar{x}$ and s are derived from a log normal plot the highs and lows for μ and σ or $\check{z}$ can be placed on the log normal plot with the original data and limits on the expectations can be made.

B. Weibull Distribution

The confidence limits for the infinite sample size Weibull curves Eqs (1.2), (1.14) and (1.15) from the test samples [1.1,1.21] are shown in Table 1.1. The test sample B and θ are estimated from a plot or a computer

Table 1.1 Confidence levels

Confidence level	$Z_{\alpha/2}$
99%	2.576
95%	1.960
90%	1.645

calculation. Then for the infinite sample size

$$B \exp\left(\frac{-0.78\, Z_{\alpha/2}}{\sqrt{N}}\right) \le \beta \le B \exp\left(\frac{0.78\, Z_{\alpha/2}}{\sqrt{N}}\right) \tag{1.37}$$

$$\theta \exp\left(\frac{-1.05\, Z_{\alpha/2}}{B\sqrt{N}}\right) \le \Theta \le \theta \exp\left(\frac{+1.05\, Z_{\alpha/2}}{B\sqrt{N}}\right) \tag{1.38}$$

These are the ranges for β and Θ. The values can be substituted into Eq. (1.16) and plotted on Weibull paper with the test data. Also note Eq. (1.15) where for infinite sample-size

$$\Theta^{\beta} = \delta$$

In terms of the test samples

$$\theta^{B} = \Delta \tag{1.39}$$

For the infinite sample size

$$\Theta = \delta^{1/\beta}$$

For Eq. (1.38) the δ for infinite sample size is in terms of Δ from raw data

$$\Delta^{1/\beta} \exp\left[\frac{-1.05 Z_{\alpha/2}}{B\sqrt{N}}\right] \le \delta \le \Delta^{1/\beta} \exp\left(\frac{1.05 Z_{\alpha/2}}{B\sqrt{N}}\right) \tag{1.40}$$

It should be noted the spread on β, θ, δ is smaller as N increases.

On Weibull paper, percent failures are plotted but Eq. (1.16) is a reliability and Eq. (1.18) is the probability of failure. Which has values when

$$0 \le (x - \gamma) \le \infty \tag{1.41}$$

or from the lowest value data point γ to the highest as partitioned using Sturges Rule. When the failure is calculated it is multiplied by 100 to get percent failure. Also note Eq. (1.18) with solutions for γ fixed at the lowest test value

$$P_f = 1 - \exp\left[\frac{-(x-\gamma)^B}{\delta}\right] = 1 - \exp\left[-\left(\frac{(x-\gamma)}{\theta}\right)^B\right] \tag{1.42}$$

Table 1.2 90% confidence bounds on the Weibull line [1.1]

Sample size (n)	$K(n)$
3	0.540
4	0.420
5	0.380
6	0.338
7	0.307
8	0.284
9	0.269
10	0.246
11	0.237
12	0.222
13	0.213
14	0.204
15	0.197
20	0.169
25	0.152
30	0.141
35	0.125
40	0.119
45	0.117
50	0.106
75	0.086
100	0.074

The plotting on Weibull paper shows the raw data. And from Table 1.2 upper and lower lines for 90% confidence may be plotted with respect to the raw data. Then a computer solution must yield a β so that the sloped line passes through the raw data and is between the confidence lines

$$P_f(R.d) - K \le P_f \le P_f(R.d) + K \tag{1.43}$$

V. GOODNESS OF FIT TESTS

The following tests [1.5,1.17,1.18,1.23] should be mentioned and judgment should be exercised as to how much information is desired.

A. Anderson–Darling Test for Normality

The MILHDBK 5F [1.8] pages 9-185 to 9-188 discusses this test which requires the calculation of the mean, $\bar{x}$, and standard deviation, s, after the raw data is processed by plotting or computer calculation. A variable is developed

$$Z_I = (x_i - \bar{x})/s \quad i = 1, \ldots, n \tag{1.44}$$

The Anderson–Darling Test, AD, statistic is

$$AD = \left[\sum_{i=1}^{n} \frac{1 - 2i}{n} \left[\ln(F_0[Z_i]) + \ln(1 - F_0[Z(n + 1 - i)]) \right] \right] - n \tag{1.45}$$

where

F_o is the area $F_o(x)$ under the Gaussian curve to the left of x

Then if

$$AD \geq 0.752[1 + 0.75/n + 2.25/n^2]^{-1} \tag{1.46}$$

The data is *not* normally distributed from the calculation for a 95% confidence level.

B. Anderson–Darling Test for Weilbullness

This [1.18] is a test for a three parameter Weibull fit of raw data and a similar variable is

$$Z_i = [(xi - \tau_{50})/\alpha_{50}]^{\beta_{50}} \quad i = 1, \ldots, n \tag{1.47}$$

However $\beta_{50}, \alpha_{50}, \tau_{50}$ require data processing. The Anderson–Darling test statistic

$$D = \left[\sum_{i=1}^{i=n} \frac{1 - 2i}{n} \left[\ln(1 - \exp[Z_i]) + \exp[Z_{(n+1-i)}] \right) \right] - n \tag{1.48}$$

if

$$AD \geq 0.757[1 + 0.2/\sqrt{n}]^{-1} \tag{1.49}$$

It is concluded the raw data is *not* part of a three parameter Weibull distribution for a 95% confidence level.

C. Qualification of Tests

When using the goodness of fit tests there is a five percent error on the test. Further the tests may reject data even when a reasonable approximation may exist in the lower tail. Then [1.8] suggests plotting the raw data for percent failures on Weibull or log normal paper. This is an integration of $f(x)$ or failure data and tends to smooth out any variations so that a better estimate of Δ, or θ, B, γ and $\bar{x}$, s can be obtained. Then Eqs. (1.37) and (1.38) can be used for the Weibull fit of test data and Eqs. (1.21) and (1.22) for the Gaussian fit of the data.

VI. PRIORITY ON PROCESSING RAW DATA

Priority is decided when looking at Sturges Rule Section 1.1 when for Example 1.1

$$K = 1 + 3.3 \log_{10} N$$
$$K = 1 + 3.3 \log_{10} 2$$
$$K = 1.993(2.00)$$

Range (R) $= x_2 - x_1 = 500$ psi

Cell width $(\omega) = \dfrac{R}{K} = \dfrac{5000\,\text{psi}}{2} = 2500$

Each cell has 1 failure and examination of probability paper for the Weibull and Gaussian distribution for percent failures may have one data point at 50% but 100% or 0% does not show on a log scale. As a result there are two plots with one data point at 50% for each and any line passes through one data point.

The estimates in Eq. (1.33)–(1.36) must be used. They are as accurate as can be obtained. The Gaussian curve data is the approximated distribution. Up to 20 test points is allowable. A computer analysis is out of the question for $N = 2$.

When $2 \leq N \leq 20$ the individual data points may be used to obtain a line on both Weibull and Gaussian distribution plots. Note in Table 1.3 individual points allow around 10 data points for $N = 10$ while the partitioning into cells allows only 4 data points rounded off. In order to obtain plotted estimates individual data may be used and fitted to partitioned cells for greater accuracy.

EXAMPLE 1.2. The rainy season annual rainfall data is published for the civic center in Los Angeles in July of each year. The data for 1877–1997 (published 4 July 1997 in the L.A. Times) is listed in Appendix F. This data, 120 values, was entered into a SAS program to generate Gaussian and the Weibull three parameter distribution. The data is arranged and partitioned according to Sturges Rule (Section I) then Fig.

Table 1.3 Cells (K) on Weibull and normal probability paper

N	K cells	
2	2	cannot obtain a line
3	2.57(3)	
4	2.99(3)	
5	3.306(3)	
6	3.56(4)	use individual data points to better
7	3.78(4)	defined lines $3 \leq N \leq 10$
8	3.98(4)	
9	4.15(4)	
10	4.30(4)	
15	4.88(5)	Sturges Rule partitioned data
20	5.29(5)	$15 \leq N \leq 100$
30	5.87(6)	
100	7.60(8)	

1.2 is plotted using Eq. (1.43) and the K values from Table 1.2. The slope $\beta = 1.516$ from Fig. 1.2 and can change as long as the line stays within the 90% confidence bounds. The SAS computer program calculates the best fit for the Gaussian distribution and iterates to find the maximum liklihood estimaters for the Weibull curve to the data Fig. 1.3. The Weibull curve, the solid line, has estimated values Eqs. (1.2), (1.14) of

$$\beta = 1.58917 \pm 0.12731 \quad \theta = 11.48905 \pm 0.76931$$
$$\gamma = 4.68106 \pm 0.23477 \text{ inches} \tag{1.50}$$

The variation Eq. (2.13) on the β is 0.01621 with $\check{z}_\beta$ Eq. (2.16) of ± 0.12731 which allows comparison with Fig. 1.2. The calculated values for the dotted line Gaussian Eq. (1.1) are for 120 data points

$$\mu = 14.97683 \quad \check{z} = 6.72417 \tag{1.51}$$

The 50th percentile from Fig. 1.2 is 11.2 inches, compared to μ of 14.97683. This means the data is skewed to the left in Fig. 1.3.

The SAS computer program calculates a goodness of fit by the following tests for normality by Anderson–Darling, Cramer–Von Mises, and Kolmogorov. The Gaussian curve is not as good of a fit as the Weibull curve which from the Weibull Anderson–Darling and Cramer–Von Mises tests is a better fit.

EXAMPLE 1.3. Aluminum casting, (24 yield strength data points) from Problem 1.1 casting A is used to obtain the best fit of a Gaussian or a Weibull distribution. A $11'' \times 17''$ plot similar to Fig. 1.2 is made

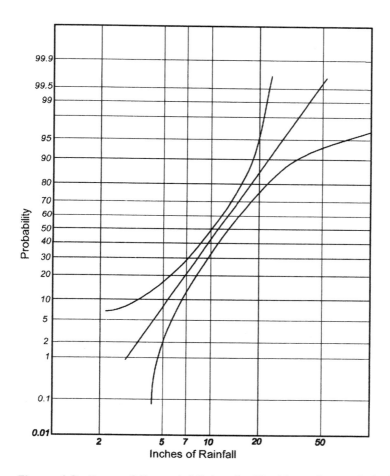

Figure 1.2 Percent failure rainfall data for Fig. 1.3 per Sturges Rule $\beta \approx 1.516$.

and the β through the data is 15.87 and the smallest β is 7.00 with the 50th percentile of 38,000–39,000 psi. The Gaussian values, Eqs. (1.1), are plotted as a dotted line in Fig. 1.4.

$$\mu = 39,370 \, \text{psi} \qquad \check{z} = 1057 \tag{1.52}$$

The solid line Weibull curve plotted in Fig. 1.4 has three parameters for Eq. (1.2), (1.14) of

$$\beta = 3.12371 \pm 0.4310 \qquad \theta = 3,218.959 \pm 1,756$$
$$\gamma = 36,497 \pm 1,634 \, \text{psi} \tag{1.53}$$

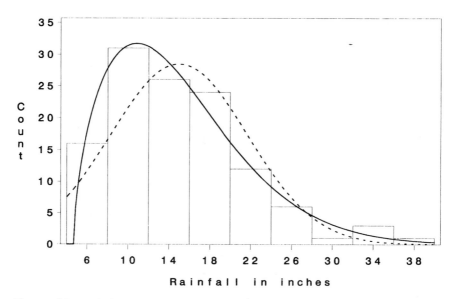

Figure 1.3 Rainfall data at civic center Los Angeles 1877–1997 for July 1–June 30.
Weibull solid line
$\beta = 1.58917 \pm 0.12731$, $\theta = 11.48905 \pm 0.76931$ $\gamma = 4.68106 \pm 0.23477$,
Gaussian dotted line
$\mu = 14.97683$, $\breve{z} = 6.72417$.

The variation Eq. (2.13) on β is 0.18572 and the $\breve{z}_\beta$ is 0.4310 (Eq. (2.16)). The
goodness of fit evaluations find both the Weibull curve and the Gaussian
curve are acceptable distributions. Visual examination of Fig. 1.4 would
justify this.

The Example 1.1 is examined for this data and one of the parameters is
the class A and B basis stress levels for design. Using Eq. (1.32) and $K = 2.25$
(Fig. E.1) for A basis and $K = 1.80$ (Fig. E.1) for B basis. $\mu = 39,370$ psi with
$s = 1057$ psi and $N = 24$.

$$\alpha_A = 39,370 - 2.25(1057\,\text{psi}) = 36,992\,\text{psi} \tag{1.54}$$

$$\alpha_B = 39,370 - 1.80\,(1057\,\text{psi}) = 37,467\,\text{psi} \tag{1.55}$$

A class C $K = 5.8$, Fig. E.1 with $N = 24$

$$\alpha_C = 39,370 - 5.80(1057\,\text{psi}) = 33,239\,\text{psi}$$

The μ or $\bar{x}$ and $\breve{z} = s$ can be corrected from $N = 24$ to infinite sample size
using Eqs. (1.30) and (1.31). The d.f. $= 24 - 1 = 23$, Table E.2 the t value

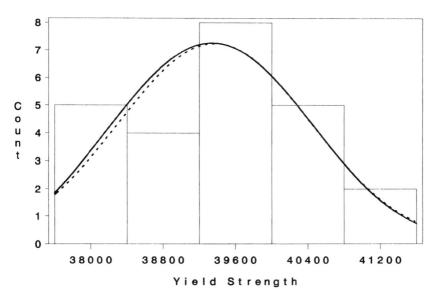

Figure 1.4 Aluminium casting (A) Problem 1.1 yield data.
Solid Weibull line
$\qquad \beta = 3.12371 \pm 0.4310 \qquad \theta = 3{,}218.959 \pm 1{,}756 \qquad \gamma = 36{,}497 \pm 1{,}634$ psi
Dotted Gaussian line
$\qquad \mu = 39{,}370$ psi $\qquad \check{z} = 1057$ psi.

for 0.025 is 2.069 and x^2 Table E.3 the 0.025 value 38.08 with square root of 6.1709 and 0.975 value 11.69 with square root value of 3.419.

Using Eq. (1.30)

$$39{,}370 - 2.069\,\frac{1057}{23} \le \mu_I \le 39{,}370 + 2.069\,\frac{1057}{23} \tag{1.56}$$
$$39{,}275 \le \mu_I \le 39{,}465$$

Using Eq. (1.31)

$$\frac{1057\sqrt{24}}{6.1709} \le \sigma_I \le \frac{1057\sqrt{24}}{3.419} \tag{1.57}$$
$$839 \le \sigma_i \le 1{,}515\,\text{psi}$$

The Weibull parameters for 24 data points, N, are Eq. (1.53).

$$\beta = B = 3.12371 \qquad \theta = 3{,}218.959 \qquad \gamma = 36{,}497\,\text{psi}$$

Convert to infinite sample size with 95% confidence using Eq. (1.37) for infinite sample size.

$$B \exp\left[\frac{-0.78z_{\alpha/2}}{\sqrt{N}}\right] \leq \beta \leq B \exp\left[\frac{0.78z_{\alpha/2}}{\sqrt{N}}\right]$$

from Table 1.1

$$z_{\alpha/2} = 1.960 \qquad N = 24$$

Substituting

$$0.731934\, B \leq \beta \leq 1.36624\, B$$
$$2.28635 \leq \beta \leq 4.26774 \tag{1.58}$$

The solution from SAS for 24 data points from Eq. (1.53)

$$2.69271 \leq B \leq 3.55471 \tag{1.59}$$

for infinite sample size Θ use Eq. (1.38)

$$\theta \exp\left[\frac{-1.05z_{\alpha/2}}{B\sqrt{N}}\right] \leq \Theta \leq \theta \exp\left[\frac{1.05z_{\alpha/2}}{B\sqrt{N}}\right]$$
$$0.874167\,\theta \leq \Theta \leq 1.14395\,\theta$$
$$2813.91 \leq \Theta \leq 3682.33 \tag{1.60}$$

The solution from SAS for 24 data points from Eq. (1.53)

$$1462.96 \leq \theta \leq 4974.96 \tag{1.61}$$

The intercept γ for 24 points has an average or mean and what appears as a standard deviation. The only option is to bound γ is by use of an α_A equation similar to Eq. (1.54) for the infinite sample size

$$\gamma_A = \bar{\gamma} \pm k_A\, S_\gamma$$
$$\gamma_A = 36{,}497 \pm 3.25(1{,}634) \tag{1.62}$$
$$31{,}187 \leq \gamma_A \leq 41{,}808\,\text{psi}$$

The SAS solution for 24 points gives from Eq. (1.53) yields

$$34{,}863 \leq \gamma \leq 38{,}131\,\text{psi} \tag{1.63}$$

Now for the infinite Weibull distribution Eq. (1.14)

β is obtained from Eq. (1.58)

Θ is Eq. (1.60) while Eq. (1.61) has more spread

γ is from Eq. (1.62)

EXAMPLE 1.4. Aluminum casting, 26 tensile strength data, from Problem 1.1 casting A is used to find the best fit of a Gaussian curve or

Weibull. A $11'' \times 17''$ plot similar to Fig. 1.2 is made and the β through the data is 6.80 and the smallest β is 2.50 with the 50 percentile of 43,500–45,500 psi. The Gaussian values Eq. (1.1) are plotted as the dotted curve in Fig. 1.5

$$\mu = 46,507 \text{ psi} \qquad \check{z} = 2158 \text{ psi} \qquad (1.64)$$

The Weibull values Eqs. (1.2), (1.14) are plotted as a solid line in Fig. 1.5

$$\beta = 1.56090 \pm 0.4316 \quad \theta = 3766.922 \pm 739 \quad \gamma = 43,108 \text{ psi} \pm 387 \text{ psi} \qquad (1.65)$$

The variation Eq. (2.13) on β is 0.1862641 and $\check{z}_\beta$ (Eq. (2.16)) is 0.4310. The best goodness of fit found is the Weibull curve.

Again using Example 1.1 and following the format of Example 1.3 the class A $K = 3.15$, Fig. E.1 and Class B $K = 1.82$, Fig. E.1 for N, 26, samples. Then with $\mu = \bar{x} = 46,507$ psi and $s = \check{z} = 2158$ psi using Eq. (1.32)

$$\alpha_A = 46,507 - 3.15(2158) = 39,709 \text{ psi} \qquad (1.66)$$

$$\alpha_B = 46,507 - 1.82(2158) = 42,579 \text{ psi} \qquad (1.67)$$

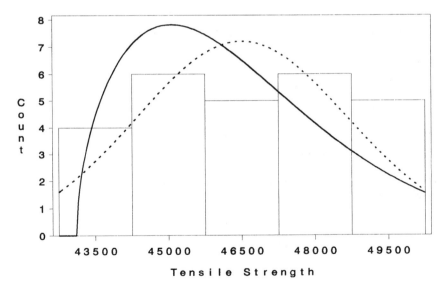

Figure 1.5 Aluminium casting (A) Problem 1.1 tensile strength.
The solid Weibull line
$\beta = 1.5609 \pm 0.4316 \qquad \theta = 3766.922 \pm 739 \qquad \gamma = 43,108 \pm 387$ psi
$\mu = 46,507$ psi $\qquad \check{z} = 2158$ psi.

A class C $K = 5.8$

$$\alpha_C = 46,507 - 5.8(2158) = 33,991 \text{ psi}$$

The μ or $\bar{x}$ and $\check{z} = s$ can be corrected from $N = 26$ to N infinite using Eqs. (1.30) and (1.31). The d.f. $= 26 - 1 = 25$, from Table E.2 the t value for 0.025 is 2.060 then x^2 from Table E.3 the 0.025 value is 40.65 with square root of 6.376 and the 0.975 value of 13.12 or $X = 3.622$.

Using Eq. (1.30)

$$46,507 - 2.060 \frac{2158}{25} \le \mu_I \le 46,507 + 2.060 \frac{2158}{25} \tag{1.68}$$

$$46,329 \le \mu_I \le 46,685 \text{ psi}$$

Using Eq. (1.31)

$$\frac{2158\sqrt{26}}{6.376} \le \sigma_I \le \frac{2158\sqrt{26}}{3.622} \tag{1.69}$$

$$1,726 \le \sigma_I \le 3,038 \text{ psi}$$

The Weibull parameter Eq. (1.65), for 26 data points will be converted from 26 points to an infinite sample size. The process starts with Eq. (1.58) and

$$z^{\alpha/2} = 1.960 \text{ Table 1.1 and } N = 26 \text{ points}$$
$$\beta = B = 1.5609 \qquad \theta = 3766.922 \qquad \gamma = 43,108$$

yields

$$0.740950 \, B \le \beta \le 1.34962 \, B$$
$$1.15655 \le \beta \le 2.10662 \tag{1.70}$$

The solution from SAS for 26 data points from Eq. (1.64)

$$1.1299 \le B \le 1.99925 \tag{1.71}$$

The Θ conversion from 26 data points to infinite sample size follows Eq. (1.60) with

$$z^{\alpha/2} = 1.960 \text{ and } N = 26 \text{ with } \theta = 3766.922$$
$$0.772152 \, \theta \le \Theta \le 1.29508 \, \theta \tag{1.72}$$
$$2908.64 \le \Theta \le 4878.47$$

The solution for 26 data points from Eq. (1.64) is

$$3027.92 \le \theta \le 4505.92 \tag{1.73}$$

The γ conversion to infinite sample size follows Eq. (1.62) with $K_A = 3.15$

from Fig. E.1 and Eq. (1.66)

$$\gamma_A = 43,108 \pm 3.15(387)$$
$$41,889 \le \gamma_A \le 44,327 \text{ psi} \tag{1.74}$$

The SAS solution for 26 points from Eq. (1.65) yields

$$42,721 \le \gamma \le 43,495 \text{ psi} \tag{1.75}$$

The infinite Weibull distribution Eq. (1.14)

β is from Eq. (170)

Θ is from Eq. (172)

γ is from Eq. (1.74)

EXAMPLE 1.5. Select the best fitting curve, Weibull, for the ultimate strength of Ti-16V-2.5Al for 755 tests [1.28]

Stress $\times 10^3$ psi	Number	Stress $\times 10^3$ psi	Number
149.6–154.05	3	176.3–180.75	181
154.05–158.5	7	180.75–185.2	148
158.5–162.95	20	185.2–189.65	47
162.95–167.4	47	189.65–194.1	20
167.4–171.85	98	194.1–198.55	5
171.85–176.3	176	198.55–203	3
			755

A $11'' \times 17''$ plot similiar to Fig. 1.2 gave $\beta = 7.7$ through the data and the smallest $\beta = 5.3$ with the 50 percentile of 175,000 psi. The Gaussian values Eq. (1.1) plotted as a dotted line in Fig. 1.6 are

$$\mu = 176.703 \text{ ksi} \qquad \breve{z} = 7.494 \text{ ksi} \tag{1.76}$$

The Weibull Eqs. (1.2) and (1.14) plotted as a solid curve in Fig. 1.6 has values for the three parameters of

$$\beta = 4.59132 \pm 0.34135 \qquad \theta = 33.850 \pm 2.234$$
$$\gamma = 145.707 \pm 2.153 \text{ ksi} \tag{1.77}$$

The variation Eq. (2.13) on β is 0.11652 and $\breve{z}_\beta$ (Eq. (2.16)) is 0.34135 The reader can follow the Examples 1.1, 1.3, 1.4 and 1.6 and can see the mean and standard deviation comparison for the sample of 755 and infinite sample size are small.

The titanium ultimate strength properties for 755 tests may be examined using Eq. (1.32) and Appendix E to find α_A, α_B, α_C one sided design stress values from a Gaussian distribution with 755 samples $K_A = 2.45$, Fig.

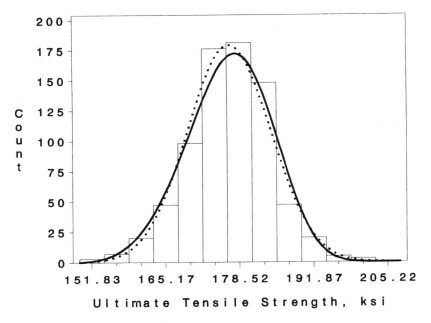

Figure 1.6   Ultimate strength, Ti−16v−2.5 Al, for 755 tests [1.28].
Solid Weibull line
    $\beta = 4.59132 \pm 0.34135$ $\theta = 33.850 \pm 2.234$ $\gamma = 145.707 \pm 2.153$ Ksi
Dotted Gaussian line
    $\mu = 176.703$ Ksi $\check{z} = 7.494$ Ksi.

E.1, $K_B = 1.35$, $K_C = 4.45$

$$\alpha_A = \bar{x} - K_A S$$
$$\alpha_A = 176.703 \text{ Ksi} - 2.45 \ (7.494 \text{ Ksi}) \tag{1.78}$$
$$\alpha_A = 158.343 \text{ Ksi}$$

$$\alpha_B = \bar{x} - K_B S$$
$$\alpha_B = 176.703 - 1.35 \ (7.494) \tag{1.79}$$
$$\alpha_B = 166.586 \text{ Ksi}$$

$$\alpha_C = 176.703 - 4.45 \ (7.494)$$
$$\alpha_C = 143.355 \text{ Ksi} \tag{1.80}$$

The infinite sample size Weibull parameters for 95% confidence from the 755 test sample using Eq. (1.37)

$$B \exp\left[\frac{-0.78\ Z_{\alpha/2}}{\sqrt{N}}\right] \leq \beta \leq B \exp\left[\frac{+0.78\ Z_{\alpha/2}}{\sqrt{N}}\right]$$

from Table 1.1

$$Z_{\alpha/2} = 1.960 \qquad N = 755$$
$$0.945881 \quad B \leq \beta \leq 1.05722\ B$$

substituting B from Eq. (1.77) the infinite sample size

$$4.34283 \leq \beta \leq 4.85402$$

while the computer yields on 755 samples

$$4.24996 \leq \beta \leq 4.93266 \tag{1.81}$$

then from Eq. (1.38)

$$\theta \exp\left[-\frac{1.05\ Z_{\alpha/2}}{B\sqrt{N}}\right] \leq \Theta \leq \theta \exp\left[\frac{1.05\ Z_{\alpha/2}}{B\sqrt{N}}\right]$$

using

$$B = 4.59132\ \text{Eq. (1.77)} \qquad N = 7.55 \qquad Z_{\alpha/2} = 1.960$$
$$0.983819\,\theta \leq \Theta \leq 1.01645\,\theta$$

with $\theta = 33.850$ Eq. (1.77)
then

$$33.3032 \leq \Theta \leq 34.4077 \tag{1.82}$$

with from the computer for 755 samples

$$31.6164 \leq \Theta \leq 36.0853$$

now the lower value of γ is evaluated using K_A values from Eq. (1.78)

$$\gamma_{AL} = x - K_A S$$
$$\gamma_{AL} = 145.707\ Ksi - 2.45\ (2.153)$$
$$\gamma_{AL} = 140.432\ \text{Ksi} \tag{1.83}$$
$$140.432\ \text{Ksi} \leq \gamma_A \leq 150.984\ \text{Ksi}$$

Also

$$136.123 \leq \gamma_C \leq 155.291\ \text{Ksi}$$

while the computer for 755 samples

$$143.554 \leq \gamma \leq 147.860\ \text{Ksi}$$

finding the mean for an infinite sample size Eq. (1.30)

$$\bar{x} - t_{0.975} \frac{S}{N-1} \le \mu_I \le \bar{x}\, t_{0.975} \frac{S}{N-1}$$

from Table E.2 d.f. $= 755 - 1 = 754$ $\qquad t_{0.975} = t_{0.025} = 1.960$

$$176.703 - 1.96 \frac{7.494}{754} \le \mu_i \le 176.703 \text{ Ksi} + 1.960 \frac{7.494}{754} \tag{1.84}$$

$$176.684 \text{ Ksi} \le \mu_i \le 176.722 \text{ Ksi}$$

The infinite standard deviation is Eq. (1.31)

$$\frac{S\sqrt{N}}{x_{0.025}} \le \sigma_I \le \frac{S\sqrt{N}}{x_{0.025}}$$

with d.f. $= N - 1$ for 150 and above [1.12]

$$\frac{x^2}{N-1} \text{ approaches 1 for 95 percentile}$$

So

$$x \approx \sqrt{N-1}$$

or

$$\sigma_I \approx s \approx 7.494 \text{ Ksi} \tag{1.85}$$

EXAMPLE 1.6. Three sets of radiator data (4.26) A, B, and C, with nine samples each.

	A	B	C
Mean	57,213 cycles	62,073	55,491
Standard Deviation	29,287 cycles	28,223	25,913
C_V	51.19%	45.47%	46.7%

In small sampling theory [1.27] the means and standard deviations may be checked for A and B also A and C so that it can be stated the samples came from a larger Gaussian or near Gaussian distribution

$\qquad H_o : \bar{x}_A = \bar{x}_B$ No difference in the two group (Eq. (1.28))

$\qquad H_i : \bar{x}_A \ne \bar{x}_B$ and there is significant difference for H_o

$$t = \frac{\bar{x}_A - \bar{x}_B}{\sigma \left[\dfrac{1}{N_A} + \dfrac{1}{N_B} \right]^{1/2}} \qquad \sigma = \left[\frac{N_A S_A^2 + N_B S_B^2}{N_A + N_B - 2} \right]^{1/2}$$

$$t = \frac{57,213 - 62,073}{\sigma\left[\frac{1}{9} + \frac{1}{9}\right]^{1/2}} \qquad \sigma = \left[\frac{9(29,287)^2 + 9(28,223)^2}{9 + 9 - 2}\right]^{1/2} \tag{1.86}$$

$$t = -0.3380 \qquad \sigma = 30,505 \text{ cycles}$$

If a significant level of 0.01 and $N_A + N_B - 2 = 16$ degrees of freedom H_o is rejected if it is outside of the range of $-t_{0.995}$ to $t_{0.995}$ where $t_{0.995}$ to ± 2.921 (Table E.2). Therefore H_o is accepted.

Now comparing A and C:

$$t = \frac{57,213 - 55,491}{29,329\left[\frac{1}{9} + \frac{1}{9}\right]^{1/2}} \qquad \sigma = \left[\frac{9(29,287)^2 + 9(25,913)^2}{9 + 9 - 2}\right]^{1/2} \tag{1.87}$$

$$t = 0.1245 \qquad \sigma = 29,329 \text{ cycles}$$

Using the same significance levels as before A and C sets are from the same larger set and so should sets AB and C.

A SAS computer run for 26 samples yields for a dotted line Gaussian distribution Fig. 1.7

$$\bar{X}_{ABC} = 60,329 \text{ cycles}$$
$$S_{ABC} = 25,145 \text{ cycles} \tag{1.88}$$

The estimate for the solid Weibull distribution Fig. 1.7

$$\beta = 1.71144 \pm 0.486265$$
$$\gamma = 18,390 \pm 6004.5 \text{ cycles} \tag{1.89}$$
$$\theta = 46,903 \pm 9671$$

The goodness of fit computer calculation finds the data fits both the Gaussian and Weibull curves.

The data is further reduced in a manner as Examples 1.3 and 1.4 following Example 1.1. The class A $K = 3.15$ Fig. E.1 and class B $K = 1.82$ Fig. E.1 for 26 sample. Then with $\bar{X}_{ABC} = 60,329$ cycles and $S_{ABC} = 25,145$ cycles Eq. (1.32) yields

$$\alpha_A = 60,329 \text{ cycles} - 3.15(25,145) = -18,878 \text{ cycles} \tag{1.90}$$

$$\alpha_B = 60,329 - 1.82\,(25,145) = 14,565 \text{ cycles} \tag{1.91}$$

The sample mean $\bar{X}_{ABC}$ and standard deviation S_{ABC} are corrected from 26 samples to infinite sample size using the same data as Example 1.4 using

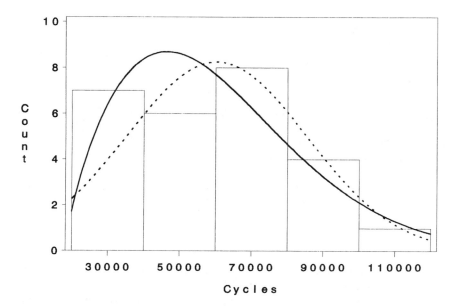

Figure 1.7 Three combined tests (A, B, C) of radiator data [4.26].
Solid Weibull curve
$$\beta = 1.71144 \pm 0.486265 \qquad \theta = 46,903 \pm 9671 \qquad \gamma = 18,390 \pm 6004.5 \text{ cycles}$$
Dotted Gaussian curve
$$\bar{x}_{ABC} = 60,329 \text{ cycles} \qquad S_{ABC} = 25,145 \text{ cycles.}$$

Eq. (1.30) then Eq. (1.31)

$$60,329 - 2.060\frac{25,145}{25} \le \mu_i \le 60,329 + 2.060\frac{25,145}{25} \tag{1.92}$$
$$58,257 \le \mu_i \le 62,401 \text{ cycles}$$

and

$$\frac{25,145\sqrt{26}}{6.376} \le \sigma_i \le \frac{25,145\sqrt{26}}{3.622} \tag{1.93}$$
$$20,109 \le \sigma_i \le 35,399 \text{ cycles}$$

REFERENCES

1.1.   Abernethy RB et al.: Weibull Analysis Handbook AFWAL-TR-2079, NTIS
        (AD-A143100) There is a 2nd edition from Gulf Publishing, 1983.

1.2.   Abernethy RB. The New Weibull Handbook, Gulf Publishing Co. 2nd ed., 1994.

1.3.   Bowker AH, Lieberman GJ. Engineering Statistics, Englewood Cliffs, NJ: Prentice-Hall, 1959, also a later 2nd ed. 1972.

1.4.   Craver JS. Graph Paper From Your Copier, Tuscon, AZ: Fisher Publishing Co, 1980.

1.5.   D'Agostina RB, Stephens MA. Goodness-of-fit techniques, Marcel Dekker, 1987, p. 123.

1.6.   Dieter GE. Engineering Design, New York: McGraw-Hill Book Co, 1983.

1.7.   Dixon JR. Design Engineering, New York: McGraw-Hill, 1966.

1.8.   Dixon WJ, Massey Jr FJ. Introduction of Statistical Analysis, 3rd ed. New York: McGraw-Hill, 1969.

1.9.   Grube KR, Williams DN, Ogden HR. Premium-Quality Aluminum Castings, DMIC Report 211, 4 Jan. 1965 Defense Metals Information Center, Battelle Institute, Columbus, OH, 1965.

1.10.   Hald A. Statistical Theory with Engineering Applications, New York: John Wiley & Sons, 1952.

1.11.   Haugen EB. Probabilistic Approaches to Design, New York: John Wiley & Sons, 1968.

1.12.   Haugen EB. Probabilistic Mechanical Design, New York: Wiley Science, 1980.

1.13.   Hogg RV, Ledolter J. Applied Statistics for Engineers and Physical Scientists, New York: MacMillian Co, 1992.

1.14.   Johnson LG. The Statistical Treatment of Fatigue Experiments, New York: Elsevier Publishing Co, 1964.

1.15.   Juvinall RC. Stress, Strain, and Strength, New York: McGraw-Hill Inc, 1967.

1.16.   King JR. Probability Charts for Decision Making, Industrial Press, 1971.

1.17.   Lawless JF. Statistical Models and Methods for Lifetime Data, John Wiley and Sons, 1982 pp. 452–460.

1.18.   Mil HDBK-5F. Metallic Materials and Elements for Aerospace Vehicle Structures, Department of Defense, 1992.

1.19.   Middendorf WH. Engineering Design, Boston: Allyn and Bacon Inc, 1969.

1.20.   Natrella MG. Experimental Statistics, National Bureau of Standards Handbook 91. August 1, 1963.

1.21.   Nelson W. Applied Life Data Analysis, New York: John Wiley and Sons, Inc, 1982.

1.22.   Owen DB. Factors for one-sided tolerance limits and for variables and sampling plans, Sandia Corporation Monograph SCR-607, March 1963.

1.23.   Pierce DA, Kopecky KK Testing goodness of fit for the distribution of errors in regression models, Biometrika, 66: 1–5, 1979.

1.24.   Salvatore D. Statistics and Econometrics, Schaums Outline, New York: McGraw-Hill, 1981.

1.25.   SAS Users Guide: Statistical Version, 6 Ed., Cary, NC: SAS Institute, Inc, 1995.

1.26.   Sines G, Waisman JL eds. Metal Fatigue, New York: McGraw-Hill Book Co, 1959.

1.27.   Spiegel MR. Statistics, 2nd ed. Schaums Outline, New York: McGraw-Hill, 1988.

1.28.   A Statistical Summary of Mechanical Property Data for Titanium Alloys, OTS-PB-161237, Battelle Memorial Institute, Columbus, Ohio, Feb. 1961.

1.29.   Vidosic JP. Elements of Design Engineering, New York: The Ronald Press Co, 1969.

1.30.   Weibull W. A statistical distribution function of wide applicability, J. Appl. Mech. 18:293–297, 1951.

1.31.   Weibull W. Fatigue Testing and the Analysis of Results, New York: Pergamon, 1961.

PROBLEMS

I. Steps for Solution for Problems

A. Discrete Data Problems 1.1–1.5 and 1.11

1.   Plot percent failures on normal and Weibull paper to estimate Parameters and a graphical check on the computer and conversions of Weibull parameters.
2.   Perform computer runs
3.   Plot bar charts with Weibull, Gaussian, and percent failure curve on a single page. Partition using Sturges Rule.
4.   Plot data and Weibull and Gaussian curves on semilog paper for reliability (1% failure) and select best representation.

B. Bar Data Problems 1.6–1.10 and 1.12

The same steps but Sturges Rules is used to calculate how many separate bars there should be.

Problem 1.1 Mechanical Evaluation Properties of an Aluminum Support Casting [1.9]

Test sample	Casting A			Casting B			Casting C		
	Tensile strength, Kpsi	Yield strength, Kpsi	Elongation %	Tensile strength Kpsi	Yield strength, Kpsi	Elongation %	Tensile strength, Kpsi	Yield strength, Kpsi	Elongation %
1	48.3	38.1	15.0	47.1	35.8	12.0	45.7	36.2	9.0
2	50.1	40.9	10.0	47.2	37.4	10.0	43.7	33.6	11.0
3	49.3	40.2	10.0	46.0	36.0	13.0	48.0	36.9	16.0
4	50.1	39.9	10.0	47.1	36.8	13.0	47.1	35.5	14.0
5	45.8	39.7	3.0	44.6	39.8	3.0	46.0	37.6	7.0
6	44.8	38.8	3.0	41.1	38.2	1.0	43.1	33.2	8.0
7	46.0	38.9	3.0	40.9	37.0	1.0	46.1	36.1	7.0
8	43.9	39.1	2.0	42.5	37.4	3.0	46.5	35.5	7.0
9	47.5	40.5	3.5	43.6	37.7	3.0	47.0	36.3	10.0
10	43.3	39.2	1.0	43.5	39.7	3.0	43.8	33.0	11.0
11	44.3	38.4	3.0	41.4	37.4	2.0	43.7	33.3	8.0
12	48.1	39.8	5.5	42.6	38.5	2.0	46.3	36.8	8.0
13	43.9	37.6	3.0	46.1	39.9	4.0	44.8	35.7	7.0
14	44.6	38.2	3.0	42.2	37.4	3.0	44.8	33.7	10.0
15	48.7	37.6	9.5	47.8	39.5	7.0	45.1	34.5	7.0
16	49.0	37.8	6.0	45.7	—	6.0	45.1	36.3	5.0
17	43.8	41.3	3.0	46.9	39.8	4.0	41.9	32.8	7.0
18	44.3	39.4	3.0	45.8	38.5	5.0	46.0	36.7	6.0
19	44.9	39.8	3.0	41.0	37.7	2.0	47.5	37.0	11.0
20	45.9	40.1	3.5	42.7	38.7	2.0	44.9	36.3	7.0
21	—	—	—	43.2	38.3	3.0	46.4	38.5	7.0
22	48.9	40.7	4.0	43.2	38.1	2.5	46.7	36.5	7.0
23	47.9	40.5	4.0	42.0	38.4	2.5	45.5	36.7	7.0
24	47.8	39.2	3.0	42.3	39.0	2.0	44.3	35.1	5.0
25	47.2	39.2	5.0	40.9	38.0	1.5	47.3	37.9	8.0
Mean	46.6	39.4	5.0	43.9	38.1	4.4	45.5	35.7	8.3

Problem 1.2 Mechanical properties of an aluminum casting alloy [1.9]

Coupon	Tensile strength, Kpsi	0.2% Offset yield strength, Kpsi	Elongation in 2 inches, %
1	53.1	44.2	5.0
2	52.6	42.8	4.0
3	52.4	43.1	5.7
4	50.4	43.8	3.6
5	52.4	44.2	6.0
6	53.6	45.1	5.5
7	51.2	41.7	4.3
8	53.8	44.3	6.4

Problem 1.3 Mechanical, properties of tens 50-T6 aluminum sand casting [1.9]

Specimen location	Tensile strength, Kpsi	0.2% Offset yield strength, Kpsi	Elongation in 2 inches, %
1	41.7	38.5	1.0
2	37.8	36.7	1.0
3	45.1	39.3	2.0
4	43.4	39.4	1.5
5	43.8	38.8	1.5
6	48.8	40.4	4.5
7	44.1	38.6	2.0
8	41.9	37.3	1.5
9	43.4	39.0	1.5
10	40.9	39.4	1.0
11	39.0	35.9	1.0
12	34.2	33.9	1.0
13	40.3	36.7	1.0
14	42.0	37.1	1.5
15	39.2	37.0	1.0
16	47.6	42.0	1.5
17	50.9	39.9	8.0
18	47.6	39.8	3.0
19	49.6	39.8	6.5
20	37.3	Not valid	1.0
21	38.2	37.8	1.0
22	49.4	39.1	5.5
23	50.2	40.0	5.0
24	51.1	40.5	6.5
25	44.2	38.3	2.0
26	42.2	38.1	1.5
27	42.7	38.0	1.5
28	42.2	38.5	1.5
29	48.8	40.8	4.0
30	48.3	40.3	3.5
31	41.8	40.6	1.0
32	43.7	36.9	2.5
33	41.4	36.6	1.5
34	46.0	39.0	2.5
35	43.5	38.5	2.0
36	39.5	37.1	1.5

Problem 1.4 Mechanical properties of an A356-T6 casting [1.9]

Specimen location	Tensile strength Kpsi	0.2% Offset yield strength, Kpsi	Elongation in 2 inches, %
1	44.9	36.0	4.5
2	44.1	35.5	4.0
3	41.4	37.4	1.5
4	49.7	35.7	10.0
5	48.0	37.4	7.0
6	47.1	36.0	5.0
7	46.6	37.5	4.5
8	45.4	35.8	4.0
9	44.3	35.6	4.0
10	45.8	36.8	3.0
11	43.7	35.3	3.0
12	43.6	35.7	3.0
13	42.2	34.9	2.5
14	45.6	35.2	7.5
15	45.5	34.3	6.5
16	47.4	38.5	5.0
17	48.8	36.9	8.0
18	45.6	36.2	5.0
19	49.3	38.0	7.5
20	49.9	36.6	7.5
21	50.6	36.8	12.0
22	51.9	39.1	13.0
23	49.5	37.9	8.0
24	39.1	34.3	2.0
25	47.2	38.8	5.0
26	50.0	38.9	10.0
27	50.1	39.1	9.0
28	47.8	38.5	6.0
29	49.8	37.4	9.0
30	50.8	37.4	10.0
31	49.2	37.2	8.0
32	49.6	37.5	10.0
33	45.5	35.7	5.0
34	45.3	37.3	4.0
35	43.3	35.3	3.0

Problem 1.5 Mechanical properties of a modified A 356 aluminum alloy sand cast 7-38 pylon [1.9]

Sample location	Tensile strength, Kpsi	0.2% Offset yield strength, Kpsi	Elongation in 2 inches, %
1	47.8	41.2	3.6
2	51.3	41.1	7.0
3	49.5	43.5	3.5
4	53.3	41.4	7.0
5	52.4	45.1	7.0
6	53.2	43.4	8.0
7	53.4	41.8	10.0
8	53.0	43.4	8.6
9	52.7	41.7	9.0
10	53.1	42.8	8.5
11	53.8	43.8	8.0
12	53.6	41.6	11.0
13	54.4	43.5	10.5
14	54.1	43.7	12.1
15	54.1	43.5	9.3
16	53.5	44.0	7.9
17	53.6	43.9	10.0
18	52.9	43.3	7.1
19	51.5	41.4	7.0
20	50.7	40.8	6.0
21	54.2	41.5	13.0
22	52.9	43.9	6.4
23	53.0	42.3	7.9
24	52.8	41.0	8.0
25	47.6	40.6	3.6

Problem 1.6 Select the best distribution to fit the following titanium Ti-8MN tensile elongation data in percent at 75°F for 116 samples [1.28]

Percent elongation	Number
8.3–10.3	4
10.3–12.3	14
12.3–14.3	6
14.3–16.3	5
16.3–18.3	22
18.3–20.3	26
20.3–22.3	26
22.3–24.3	10
24.3–26.3	3

Problem 1.7 Select the best distribution to fit the following titanium Ti-16V-2.5Al yield data for solution treated and aged at 75°F [1.28] for 130 samples

Yield strength Kpsi	Number
142.95–148.80	1
148.80–154.65	4
154.65–160.50	11
160.50–166.35	35
166.35–172.20	21
172.20–178.05	11
178.05–183.90	18
183.90–189.75	22
189.75–195.60	7

Problem 1.8 Select the best distribution to fit the following titanium Ti-6A1-4V tensile modules of elasticity data for 115 samples [1.28]

Modulus $\times 10^6$ psi	Number
11.45–12.1	1
12.1–12.75	1
12.75–13.4	10
13.4–14.05	32
14.05–14.7	29
14.7–15.35	13
15.35–16.00	2
16.00–16.65	16
16.65–17.3	6
17.3–17.95	5

Problem 1.9 Select the best distribution for titanium Ti-5A1-2.5SN tensile elongation in percent for 75°F [1.28] for 1835 samples

Percent Elongation	Number
6.8–7.9	16
7.9–9.0	34
9.0–10.1	108
10.1–11.2	137
11.2–12.3	255
12.3–13.4	296
13.4–14.5	363
14.5–15.6	414
15.6–16.7	134
16.7–17.8	60
17.8–18.9	12
18.9–20.0	6

Problem 1.10 The life experience of electric lamps[1]
Select the best distribution for the data

Call no.	Class interval	Renewals of original units, or frequency
0	0–99.5	755
1	99.5–199.5	1,142
2	199.5–299.5	2,340
3	299.5–399.5	3,322
4	399.5–499.5	3,775
5	499.5–599.5	4,303
6	599.5–699.5	4,983
7	699.5–799.5	5,511
8	799.5–899.5	5,738
9	899.5–999.5	5,888
10	999.5–1,099.5	5,888
11	1,099.5–1,199.5	5,838
12	1,199.5–1,299.5	5,511
13	1,299.5–1,399.5	4,983
14	1,399.5–1,499.5	4,303
15	1,499.4–1,599.5	3,775
16	1,599.5–1,799.5	3,322
17	1,699.5–1,799.5	2,340
18	1,799.5–1,899.5	1,142
19	1,899.5–1,999.5	755
		75,614

[1] Taken from the mortality experience of electric lamps obtained from periodic inspections of lamps by the National Electric Lamp Association in 1915. The mortality is due entirely to use and therefore does not represent any replacements due to inadequacy, obsolescence or public requirements.

Problem 1.11 Tensile strength of steel bolts (pounds)

Producer A	Producer B	Producer C
9,220	9,930	9,570
10,030	10,040	9,280
9,180	9,850	9,350
9,250	9,730	9,430
10,150	9,330	9,710
9,330	9,890	9,570
9,090	10,100	9,750
8,910	9,330	9,310
9,140	9,670	9,140
9,230	9,590	9,640
9,310	9,240	9,670
9,230	9,540	9,010
10,200	9,160	9,180

Do the three sets of data come from a Gaussian distribution? Find the best distribution for the separate and combined data.

Problem 1.12 Data for 161 tests are taken on the coefficient of friction between two surfaces

μ	Number
0.13	1
0.14	1
0.15	6
0.16	84
0.17	17
0.18	19
0.19	2
0.20	11
0.21	7
0.22	0
0.23	5
0.24	1
0.25	3
0.26	0
0.27	4

What value of μ, coefficient of friction, would you use for designs and why? Note how $1 - P_f$ (Eq. (1.17)) which is $R(\mu)$ is close to $R(\mu) = \text{Exp}(-\lambda\mu)$ a form used in Chapter 4.

2
Application of Probability to Mechanical Design

I. PROBABILITY

Probability, to anyone who deals with concrete ideas of whether a part or system will function, is not an exact science. Even after a probability is calculated it is not certain exactly what information the individual has to work with or needs. We start our discussion with

$$P(A) = \lim_{N \to \infty} \frac{n_A}{N} \qquad 0 \leq P(A) \leq 1 \tag{2.1}$$

Now $P(A)$ is probability of A even occurring
  N is total number of events in which A can be the outcome
  n_A is the number of events in which A is the outcome.

$P(A)$ can also be thought of as an archery target area where the small area n_A is the bullseye (Fig. 2.1) and represents the area where A happens or

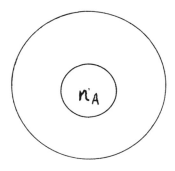

Figure 2.1 Probability of hitting a bullseye.

arrows in the bullseye and the large area is N or number of arrows shot hence the total number of events. For probability problems where two events or more occur there is more complexity. Take two events A and B

$$P(A + B) = P(A) = P(B) - P(AB) \qquad (2.2)$$

$P(A + B)$ means either A or B can happen or both and $P(AB)$ is the probability A happens followed by B. In terms of areas Eq. (2.2) is shown in Fig. 2.2.

EXAMPLE 2.1 [2.7]. The chance of success of a moon rocket is 0.20. What is the probability of success (Eq. (2.2)) if two rockets are sent.

$P(A + B) = P(A) + P(B) - P(AB)$

$P(A) = 0.20$

$P(B) = 0.20$

$P(AB) = P(A)\,P(B)$ for events which happen independently.

In other words shot A can succeed, independently of shot B.

$P(A + B) = 0.20 + 0.20 - 0.04 = 0.36$

EXAMPLE 2.2. Consider the possibility of drawing an ace (A) or any spade (B) from a full deck of cards.

$P(A + B) = P(A) + P(B) - P(AB) =$ probability of drawing an ace or
a spade or both

$P(A) = 4$ aces/52 cards

$P(B) = 13$ spades/52 cards

$P(AB)$ is one card being the ace of spade

$$P(A + B) = \frac{4}{52} + \frac{13}{52} - \frac{1}{52} = \frac{16}{52} = \frac{4 \text{ chances}}{13 \text{ trys}}$$

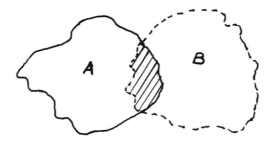

Figure 2.2 Probability of overlapping of events A and B.

For a case of three events $A + B + C$ let $X = A' + B'$ for A in Eq. (2.2) then

$$P(A + B + C) = P(X + C) = P(X) + P(C) - P(XC)$$
$$= P(X) + P(C) - P(X)P(C)$$

now substitute

$$X = A' + B'$$
$$P(x) = P(A' + B') + P(C) - P(C)[P(A' + B')]$$

now use Eq. (2.2) for $P(A' + B')$

$$P(A + B + C) = P(A') + P(B') + P(C) - P(A'B') - P(C)[P(A')$$
$$+ P(B') - P(A'B')]$$

Also it is found

$P(A'B) = P(A')P(B')$ for independent events which is substituted into the equation

$$P(A + B + C) = [P(A') + P(B') - P(A')P(B')] + P(C)$$
$$- P(C)][P(A') + P(B') - P(A')P(B')]$$
$$= P(A') + P(B') + P(C) - P(A')P(B') - P(A')P(C')$$
$$- P(C)P(B') + P(A')P(B')P(C)$$

dropping the primes yields

$$P(A + B + C) = P(A) + P(B) + P(C) - P(A)P(B) - P(A)P(C)$$
$$- P(C)P(B) + P(A)P(B)P(C) \tag{2.3}$$

EXAMPLE 2.3 [2.17]. Discrete events E1 or E2 may be approached using Example 2.2, ($\bar{E}_1$ or $\bar{E}_2$ means it doesn't occur).

n1	n2	n3	n4	n
$E_1\ E_2$	$E_1\ \bar{E_2}$	$\bar{E_1}\ E_2$	$\bar{E_1}\ \bar{E_2}$	Total

$$P(E_1) = \frac{n_1 + n_2}{n} \qquad P(E_2) = \frac{n_1 + n_3}{n}$$

Now consider $P(E_1 + E_2)$ probability of E_1 or E_2 or both

$$P(E_1 + E_2) = \frac{n_1 + n_2 + n_3}{n}$$

Now to substitute for $P(E_1)$ and $P(E_2)$ which are sums yielding

$$P(E_1 + E_2) = P(E_1) + P(E_2) - P(E_1 E_2) = \frac{2n_1 + n_2 + n_3}{n} - \frac{n_1}{n}$$

$\dfrac{n_1}{n}$ which is $P(E_1, E_2)$ probability of E_1 followed by E_2. We have

$$P(E_1 + E_2) = P(E_1) + P(E_2) - P(E_1, E_2) \text{ again Eq. (2.2).}$$

Before proceeding to Bayes theorem it is known

$$P(A) + P(\bar{A}) = 1 \tag{2.4}$$

Lets take the rocket shots Example 2.1

$P(A) = 0.20$ probability of success
$P(\bar{A}) = 1 - 0.20 = 0.80$ probability of failure.

II. BAYES THEOREM

Lets now consider two events which are not independent or

$$P(AB) = P(A)P(B/A) \tag{2.5}$$

also

$$P(BA) = P(B)P(A/B) \tag{2.6}$$

using Eqs. (2.5) and (2.6)

$$P(A/B) + \frac{P(A)P(B/A)}{P(B)} \tag{2.7}$$

Some definitions are in order.

$P(AB)$ – The probability that "A" happened followed by B.
$P(A)$ – It is not known whether or not "B" happened. $P(A)$ is the probability that "A" did.
$P(B)$ – It is not known whether or not "A" happened. $P(B)$ is the probability that "B" did.
$P(B/A)$ – "A" is known to have happened. This is the probability that it was followed by "B".
$P(A/B)$ – "B" is known to have happened. This is the probability that it was followed by "A".

Now noting Eq. (2.4)

$$P(A) + P(\bar{A}) = 1$$

B must occur with A or $\bar{A}$ so

$$P(B) = P(\bar{A}B) + P(\bar{A}B)$$

and

$$P(B) = P(A)P(B/A) + P(\bar{A})P(B/\bar{A}) \tag{2.8}$$

placing Eq. (2.8) into Eq. (2.7)

$$P(A/B) = \frac{P(A)P(B/A)}{P(A)P(B/A) + P(\bar{A})P(B/\bar{A})} \tag{2.9}$$

for "A" with more than two alternates

$$P(A/B) = \frac{P(A)P(B/A)}{\sum_i P(A_i)P(B/A_i)} \tag{2.10}$$

EXAMPLE 2.4. [2.17]. Using the Example 2.3 table for E_1 and E_2 find $P(E_2/E_1)$ using Eq. (2.5). The probability of "E_1" has happened, that "E_2" will follow, note "E_2" happens in n_1 and n_3 but "E_1" only occurs in n_1 so

$$P(E_2/E_1) = \frac{n_1}{n_1 + n_2} = \frac{\dfrac{n_1}{n}}{\dfrac{n_1 + n_2}{n}} = \frac{P(E_1 E_2)}{P(E_1)}$$

should E_1 and E_2 be independent which means E_1 and E_2 can happen separately or it means when E_1 occurs E_2 does not follow.

$$P(E_1 E_2) = P(E_1)P(E_2)$$

now

$$P(E_1) = \frac{n_1 + n_2}{n} = 2/4$$

$$P(E_2) = \frac{n_1 + n_3}{n} = 2/4$$

$$P(E_1 E_2) = (2/4)(2/4) = \frac{4}{16} = \frac{1}{4} \text{ or } \frac{n_1}{n}$$

if the events are not independent

$$P(E_1 E_2) = P(E_2)P(E_2/E_1)$$

EXAMPLE 2.5. A sorting example is solved using Eq. (2.8).

Given are two urns in a box, urn 1 with 3 white balls and 5 red balls; urn 2 with 5 white balls and 7 red balls.

What is the probability of picking a red ball without looking from this set up. Use Eq. (2.8).

$$P(B) = P(A)P(B/A) + P(\bar{A})P(B/\bar{A})$$

Let

$B = $ red ball

$A = $ urn 1

$\bar{A} = $ urn 2 since its not urn 1

Now

$$P(1) = P(2) = \frac{1}{2} \qquad P(R/1) = \frac{5}{8} \qquad P(R/2) = \frac{7}{12}$$

So the sum of the joint probabilities are

$$P(R) = P(1)P(R/1) + P(2)P(R/2)$$

$$P(R) = \left(\frac{1}{2}\right)\left(\frac{5}{8}\right) + \frac{1}{2}\left(\frac{7}{12}\right)$$

$$P(R) = \frac{5}{16} + \frac{7}{24} = \frac{15 + 14}{48} = \frac{29}{48}$$

III. DECISION TREES [2.33]

A means of analyzing logical possibilities is a decision tree and to demonstrate, Example 2.5 is reworked. The probability of picking a red ball out of a box with urn 1 with 3 white balls, 5 red balls and urn 2 with 5 white balls and 7 red balls

EXAMPLE 2.6.   In the diagram urn selection is an even option (1/2) and once an urn is selected the R or W selection is the ratio of the balls in each urn. The second draw following the flow diagram the R or W ratio is reduced by one if a R or W ball is selected and so is the total number in the urn for the second draw.

In the first draw, the probability of drawing a red ball is in Fig. 2.3.

$$P(R) = 1/2(5/8) + 1/2\left(\frac{7}{12}\right) = \frac{29}{48}$$

The same reasoning could be used to select parts out of vendor's boxes as to how many defective parts could be selected for an assembly. Decision trees by Tribus [2.55] are used to decide on testing or reworking parts during

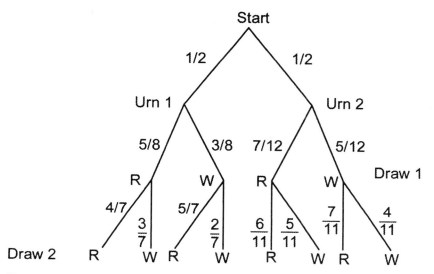

Figure 2.3 Successive draws from a supply box.

assembly and the associated costs. The following example depicts some of the concerns during manufacturing assembly.

EXAMPLE 2.7 [2.39]. An automatic assembly machine positions board A on top of board B, both nominally one inch thick with a properly drilled hole, and a machine screw, nominally $2\frac{1}{4}$ inches long, with a controlled $\frac{1}{4}$ inch thick nut torqued on at final assembly. Upon assembly for boards A and B the machine also selects an oversized $1\frac{1}{8}$ inch board 1 in 100 times, and undersized $2\frac{1}{8}$ inch screws C are selected 1 in 50 times.

The nuts have a sorter to maintain $\frac{1}{4}$ inches thickness, and, the drill holes are also controlled.

There are three assembly failures

AB–$2(1\frac{1}{8}''$ over sized) and nut can't be started on $2\frac{1}{4}$ inch screw.
BC–$B(1\frac{1}{8}''$ over sized) and C-$2\frac{1}{8}''$ allows half engagement of nut.
CA–$A(1\frac{1}{8}''$ over) and C-$2\frac{1}{8}''$ allows half engagement of nut.

Therefore how many failures may occur.

Then $P(A) = \dfrac{1}{100} = P(B)$ with $P(C) = 1/50$. Make the substitution

$$x = AB \qquad y = BC \qquad z = CA$$

$$P(AB + BC + CA) = P(x + y + z)$$

Using Eq. (2.3) with proper substitution

$$P(x + y + z) = P(x) + P(y) + P(z) - P(x)P(y) - P(x)P(z)$$
$$- P(z)P(y) + P(xyz)$$

Note combination

$P(x)P(y) - AB$ followed BC which can't happen since ABC are controlled 1 each for an assembly. Therefore only the first three conditions can occur.

$$P(x) = P(AB) = P(A)P(B) = \left(\frac{1}{100}\right)\left(\frac{1}{100}\right)$$

$$P(y) = P(BC) + P(B)P(C) = \left(\frac{1}{100}\right)\left(\frac{1}{50}\right)$$

$$P(z) = P(CA) = P(C)P(A) = \left(\frac{1}{50}\right)\left(\frac{1}{100}\right)$$

$$P(AB + BC + CA) = \left(\frac{1}{100}\right)\left(\frac{1}{100}\right) + \left(\frac{1}{100}\right)\left(\frac{1}{50}\right)$$

$$+ \left(\frac{1}{100}\right)\left(\frac{1}{50}\right) = \frac{1}{2000} \frac{\text{failures}}{\text{assemblies}}$$

A solution is to sort boards A and B and screw C. However, the cost of one board sorter and one screw sorter must be compared to the cost of ending 1/2000 failures. It should be noted that a nut sorter as well as a check on the drilled holes can be factored into the assembly.

Should the nut be oversized for 1/50 this would add another possible failure mode.

There are situations when events can happen in several different ways then the permutations and combinations must be examined.

EXAMPLE 2.8.   Find the probability that of 5 cards drawn from a deck, two will be aces. Proceed knowing the aces can be drawn in several ways, in fact, the permutations are for 5 cards, n with two of them aces, r.

$$C\binom{n}{r} = \frac{n!}{r!(n-r)!} = \left[\frac{1 \cdot 2 \cdot 3 \cdot 4 \cdot 5}{(1 \cdot 2)(1 \cdot 2 \cdot 3)}\right] = 10 \tag{2.11}$$

So

$$P\left(\begin{array}{c} 2 \ of \ 5 \ cards \\ being \ aces \end{array}\right) = \sum_{i=1}^{i=10} P \text{ (any one arrangement)}$$

assuming each has the same probability regardless of arrangement

$$P(2A \text{ in } 5 \text{ card}) = 10P(AANNN)$$

$$P(AANNN) = P(A)P(A/A)P(N/AA)P(N/AAN)P(N/AANN)$$

$$P(A) = 4/52$$

$$P(A/A) = 3/51 \quad \text{with three aces still in the deck}$$

$$P(N/A, A) = \frac{48}{50} \quad \text{any card minus two aces}$$

$$P(N/A, A, N) = \frac{47}{49}$$

$$P(N/AANN) = \frac{46}{48}$$

$$P(AANNN) = 10\left(\frac{4}{52}\right)\left(\frac{3}{51}\right)\left(\frac{48}{50}\right)\left(\frac{47}{49}\right)\left(\frac{46}{48}\right)$$

$$= [\approx 0.0399 \quad (1 \text{ chance/25 tries})]$$

IV. VARIANCE

A. Total Differential Estimate of the Variance

In this discussion some of the statements from [2.18–2.24] are stated without proof and it should be noted sample sizes are infinite. A function ψ $(x, y, z, \ldots)$ with a total differential of

$$\psi_i - \bar{\psi} = \delta\psi_i = \frac{\partial\psi}{\partial x}dx + \frac{\partial\psi}{\partial y}dy + \frac{\partial\psi}{\partial z}dz \ldots \tag{2.12}$$

Has the estimate of the variance $\check{Z}^2_\psi$ as

$$\check{Z}^2_\psi = \frac{\sum(\delta\psi)^2}{n} \tag{2.13}$$

$$(\delta\psi)^2 = \left(\frac{\partial\psi}{\partial x}\right)^2(\delta x_i)^2 + 2\frac{\partial\psi}{\partial x}\frac{\partial\psi}{\partial y}\delta x_i\delta y_i + \left(\frac{\partial\psi}{\partial y}\right)^2(\delta y_i)^2 \tag{2.14}$$

$$\check{Z}^2_\psi = \left(\frac{\partial\psi}{\partial x}\right)^2\frac{\sum(\delta x_i)^2}{n} + 2\frac{\partial\psi}{\partial x}\frac{\partial\psi}{\partial y}\frac{\sum(\delta x_i\delta y_i)}{n} + \left(\frac{\partial\psi}{\partial y}\right)\frac{\sum(\delta y_i)^2}{n} \tag{2.15}$$

if x_i and y_i are independent

$$\sum(\delta x_i\delta y_i) = 0$$

with

$$\check{Z}_x^2 = \frac{\sum(\delta x_i)^2}{n} \quad \text{and} \quad \check{Z}_y^2 = \frac{\sum(\delta y_i)^2}{n} \quad \text{the } approximate$$

standard deviation for a function is

$$\check{Z}_\psi \approx \left[\sum_{j=1}^{j}\left(\frac{\partial\psi}{\partial x_j}\right)^2 \check{z}_{xj}^2\right]^{1/2} \tag{2.16}$$

where $\check{z}_{xj}$ is standard deviation of each independent variable.

Given a normal or Gaussian function for the sum of variables x and y

$$z = x \pm y \tag{2.17}$$

where x and y are distributed normally with

$$\frac{\partial z}{\partial x} = \frac{\partial z}{\partial y} = 1$$

and $\check{z}_x$ and $\check{z}_y$ are standard deviations of x and y substituted into Eq. (2.16)

$$\check{z}_z^2 = \left[\left(\frac{\partial z}{\partial x}\right)^2 \check{z}_x^2 + \left(\frac{\partial z}{\partial y}\right)^2 \check{z}_y^2\right] \tag{2.18}$$

$$\check{z}_z^2 = \check{z}_x^2 + \check{z}_y^2$$

The Gaussian function for the product of variables x and y

$$z = xy \tag{2.19}$$

where

$$\frac{\partial z}{\partial x} = y, \qquad \frac{\partial z}{\partial y} = x$$

with the means μ_x and μ_y substituted with $\check{z}_x$ and $\check{z}_y$ the standard deviations into Eq. (2.16)

$$\check{z}_z^2 = \mu_y^2\check{z}_x^2 + \mu_x^2\check{z}_y^2 \tag{2.20}$$

The division of variables x and y

$$z = \frac{x}{y} \tag{2.21}$$

where

$$\frac{\partial z}{\partial x} = \frac{1}{y}; \qquad \frac{\partial z}{\partial y} = -\frac{x}{y^2}$$

and again the means substituted for x and y and $\check{z}_x$ and $\check{z}_y$ substituted into

Eq. (2.16)

$$\breve{z}_z^2 = \frac{1}{\mu_y^2}\breve{z}_x^2 + \frac{\mu_x^2}{\mu_y^4}\breve{z}_y^2$$

or

$$\breve{z}_z^2 = \frac{\mu_y^2\breve{z}_x^2 + \mu_x^2\breve{z}_y^2}{\mu_y^4} \tag{2.22}$$

The derivation Eqs. (2.12)–(2.16) is for what is termed *uncorrelated variables*. This means all the variables in the equation are independent of each other and a single variable can be changed without changing the value of the rest. This may be seen when examining Eq. (2.12). The case of E, Youngs modulus, from a tension test is not a result of uncorrelated variables where the

$$E = \frac{\sigma}{\in} \tag{2.23}$$

stress in the sample is divided by the strain. Hence stress or strain can not be varied independently of each other. The Eq. (2.23) consists of a measure of the force applied and the elongation because of it making Young's Modulus Eq. (2.23) a correlated variable. The mean and standard deviations are discussed by Haugen [2.18] and Miscke [2.42] for both correlated and uncorrelated variables.

It should also be noted that many of the terms like "E" and "σ" are quoted as uncorrelated variables. The relationship for coefficient of variation is developed which is the Gaussian standard deviation divided by its mean and multiplied by 100 for

$$C_v = \frac{\breve{z}}{\mu} \times 100 \quad \text{a percentage.} \tag{2.24}$$

which gives a percentage variation which becomes a constant number for various materials. Some of the values quoted in the literature and [2.18,2.19,2.44,2.53] are shown in Table 2.1. Haugen [2.18] performed an extensive study of many design parameters.

EXAMPLE 2.9. When parts are placed in an assembly the overall average assembly dimension and its variation are important. The problem of the stack up variation in parts can be examined using Eq. (2.18) and the stack up of $z = x + y$. If the coefficient of variation for x and y are

$$C_{vx} = \frac{\breve{z}_x}{\mu_x} = \pm\frac{0.01}{2.576} \quad C_{vy} = \frac{z_y}{\mu_y} = \pm\frac{0.01}{2.576}$$

The dimensions vary 1% about the means $\pm 2.576\,\breve{z}$ approximately for 99%

Table 2.1 Typical examples of coefficient of variation percentages reported in the literature

Condition	Cv percent
Concrete beam strength	15
Weld strength	10
Buckling strength thin wall cylinders	20
Timber flexure strength	16
Tensile strength of metallics	5
Yield strength of metallics	7
Tensile strengths of filamentary composites	12
Endurance limit for steel	8
Aluminum and steel modulus	3
Titanium modulus	9

of the data. Substituting into Eq. (2.18)

$$\check{z}_z^2 = \left(\frac{0.01}{2.576}\mu_x\right)^2 + \left(\frac{0.01}{2.576}\mu_y\right)^2$$

which simplifies to

$$\check{z}_z^2 \pm \frac{0.01}{2.576}[\mu_x^2 + \mu_y^2]^{1/2}$$

To obtain C_{vz}

$$C_{vz} = \frac{\check{z}_z}{x+y} = \pm\frac{0.01}{2.576}\frac{[\mu_x^2 + \mu_y^2]^{1/2}}{\mu_x + \mu_y}$$

expanded to seven stack up dimension variables

$$z = \sum_{i=1}^{i=7} x_i$$

the percent coefficient of variation for z

$$C_{vz} = \frac{\check{Z}_z}{Z} \times 100 = \pm\frac{0.01}{2.576}\frac{\left[\sum\limits_{I=1}^{I=7}\mu_{XI}^2\right]^{1/2}}{\sum\limits_{I=1}^{I=7}\mu_{XI}} \times 100$$

$$C_{vz} = \pm\frac{0.01}{2.576}\frac{11.83222}{28} \times 100$$

$$C_{vz} = \pm 0.16404\%$$

The end dimension will be

$$\mu_z = 28 \text{ in}$$

$$\breve{z}_z = C_{vz} \frac{\sum \mu_{xi}}{100}$$

$$\breve{z}_z = 0.16404 \left(\frac{28}{100} \right)$$

$$\breve{z}_z = 0.0459325$$

the stack up dimension will be

$$z = \sum_{i=1}^{i=7} x_i \pm 2.576(0.0459325)$$

$$z = 28 \text{ in} \pm 0.11832 \text{ in}$$

for 99% of the assembly.

B. Card Sort Solution Estimate of Variance

The normal functions for z examined have nicely behaved partial derivatives. However there are functions which can take several pages of partial derivatives to evaluate.

Therefore it becomes time consuming to obtain an answer when a close estimate might suffice. Further using a computer to generate distributions may also be too time consuming. The card sort selects specific values of the variables to find either or

$$z_{max} - \mu_z = X\breve{z}_c \tag{2.25}$$

$$\mu_z - z_{min} = X\breve{z}_c \tag{2.26}$$

the cards or variables are selected to make z a maximum or minimum. If the variable appears in the denominator (bottom) a large value card makes z grow toward a minimum while a small value card makes z grow toward a maximum. If a variable appears in both the denominator and numerator (top) the z_{max} means the large value card is used both places since only one card or value is used in both places. When more complicated functions are used the functions or combinations will have to be examined to see if the cards selected cooperate to yield z_{max} or z_{min}.

The variables are between two bounds for 99% for approximately ± 2.576 standard deviations of the data. The probability of being greater or less than $(x_i)_{max}$ or $(x_i)_{min}$ is $0.01/2$.

When a function z is maximized several cards or variables are select so probability of

$$P(z) \geq P(z_{max}) \tag{2.27}$$

and this is a card selection of variables separately

$$P(z) \geq P(z_{max}) \geq P(x_1 x_2 x_3 x_4 \ldots x_n) \tag{2.28}$$

from Example 2.1

$$P(x_1 x_2 x_3 x_4 \ldots x_n) = P(x_1)P(x_2)P(x_3)P(x_4) \ldots P(x_n) \tag{2.29}$$

$$P(x_1) = P(X_i) = \frac{0.01}{2} \tag{2.30}$$

So Eq. (2.29)

$$P(x_1 x_2 x_3 x_4 \ldots x_n) = \left(\frac{0.01}{2}\right)^n \tag{2.31}$$

This is the area under the Gaussian tail beyond zmax or below zmin. This one sided Gaussian curve (Fig. 2.4) and Table 2.2 may be analyzed to see how many standard deviations, $X\check{z}_c$, this range Eq. (2.25), Eq. (2.26) represents.

EXAMPLE 2.10. Example 2.9 is now solved using a card sort method.

Eq. (2.25) is set up for

$$z = x + y$$

Table 2.2 Tabulated values for $(P(0.01/2))^n$ and X for Fig. 2.4

n	$-X$	n	$-X$
1	2.5758	9	9.4351
2	4.0556	10	9.9754
3	5.1577	11	10.488
4	6.0737	12	10.9779
5	6.8738	13	11.4467
6	7.593	14	11.8974
7	8.2516	15	12.3318
8	8.8627	16	12.7517

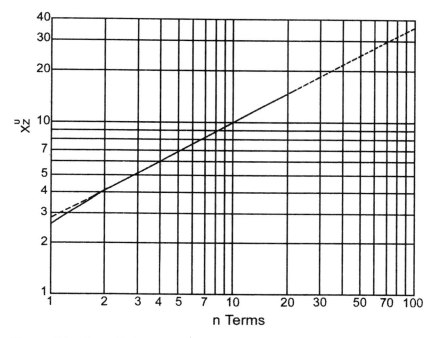

Figure 2.4 One sided standard deviation, $\bar{x}$ and $\check{z}$ for n cards selection for a function of n variables for $P_f = \left(\frac{0.01}{2}\right)^n$.

with the same conditions for a single card

$$x_{max} = 2.576\left(+\frac{0.01}{2.576}\right)\mu_x + \mu_x$$
$$x_{max} = 1.01\ \mu_x$$

the same expression holds for y_{max} and a second card

$$z_{max} = 1.01\ \mu_x + 1.01\ \mu_y$$

then

$$\mu_z = \mu_x + \mu_y$$

when Fig. 2.4 and Table 2.2 are examined for n cards (and in this case $n = 2$),

$X\breve{z}_c$ is 4.056 z_c so that

$$X\breve{z}_c = z_{max} - \mu_z$$
$$4.056 \, \breve{z}_c = 0.01 \, \mu_x + 0.01 \, \mu_y$$
$$\breve{z}_c = \frac{0.01 \, \mu_x + 0.01 \, \mu_y}{4.056}$$

and

$$C_{vz} = \frac{\breve{z}_c}{\mu_z} = \frac{0.01(\mu_x + \mu_y)}{4.056(\mu_x + \mu_y)}(100) = \frac{0.01}{4.056}(100)$$
$$C_z = 0.24654\%$$

for the second part of Example 2.9 and expanding for seven cards with $X = 8.2516$ Fig. 2.4 and Table 2.2

$$C_{vz} = \frac{\breve{z}_c}{\mu_z} = \frac{0.01}{8.2516} \frac{\sum \mu_{xi}}{\sum \mu_{xi}}(100)$$
$$C_v = 0.12119 \text{ percent}$$

The stack up dimension is $\mu_z = 28$ in

$$\breve{z}_c = \frac{0.12119}{100}(28) = 0.03393$$

the stack up dimension will be

$$z = \sum_{i=1}^{i=7} \mu_i \pm 2.576(0.03393)$$
$$z = 28 \text{ in} \pm 0.087404 \text{ in}$$

the percent error is in the tolerance

$$\text{percent error} = \frac{0.11832 - 0.087404}{0.11832} \times 100$$
$$= 26.13\% \text{ on the low side.}$$

Example 2.9 is considered to be the exact solution.

EXAMPLE 2.11.   Find the R_T for two 10% resistors in parallel (Fig. 2.5) whose mean and standard deviations are $(\mu_1, \breve{z}_1) = (10,000; 300)$ ohms and $(\mu_2, \breve{z}_2) = (20,000; 600)$ ohms. Use Eq. (2.16). R_T is

$$\frac{1}{R_T} = \frac{1}{R_1} + \frac{1}{R_2} = \frac{R_1 + R_2}{R_1 R_2}$$

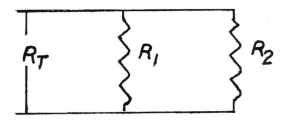

Figure 2.5 Parallel resistors for Example 2.11.

R_T is

$$R_T = F(R_1 R_2, R_1 + R_2)$$

Substituting the mean values μ_1 and μ_2

$$\mu_T = 6667 \text{ ohms}$$

the standard deviation is Eq. (2.16)

$$\check{z}_T \left[\sum_{i=1}^{i=1} \left(\frac{\partial R_T}{R_i} \right)^2 (\check{z}_i)^2 \right]^{1/2}$$

Figure 2.5 Parallel Resistors for Example 2.11 taking partial derivatives of R_T with respect to R_1 and R_2 then substituting the mean values yields

$$\frac{\partial R_T}{\partial R_1} = \frac{\mu_2^2}{(\mu_1 + \mu_2)^2} = 0.444$$

$$\frac{\partial R_T}{\partial R_2} = \frac{\mu_1^2}{(\mu_1 + \mu_2)^2} = 0.1111$$

substituting into the $\check{z}_T = [(0.4444[300])^2 + (0.1111[600])^2]^{1/2}$

$$\check{z}_T = \pm 149.1 \text{ ohms}$$

therefore

$$(\mu_T, \check{z}_T) = 6667 \pm 149.1 \text{ ohms}$$

This is considered to be an exact solution for the standard deviation. The coefficient of variation is

$$C_T = \frac{\check{z}_T}{\mu_T} 100 = 2.24\%$$

EXAMPLE 2.12. Examine Example 2.11 using a card sort to obtain the standard deviation $\check{z}_T$. In Example 2.11

$$\mu_T = 6667 \text{ ohms}$$

and

$$\check{z}_1 = 300 \text{ ohms with } \check{z}_2 = 600$$

in order to obtain $R_{T\,min}$ multiply $\check{z}_1$ and $\check{z}_2$ by 2.576 standard deviations and subtract from μ_1 and μ_2

$$R_{T_{min}} = \frac{R_{1\,min} R_{2\,min}}{R_{1\,min} + R_{2\,min}} = \frac{(9227)(18,454)}{(9227) + (18,454)} = 6151 \text{ ohms.}$$

From Eq. (2.26) and from Fig. 2.4 and Table 2.2 for two card sorts $X = 4.056$.

$$\mu_T - R_{T_{min}} = X z_c$$

$$\check{z}_c = \frac{6667 - 6151}{4.056} = 127.22 \text{ ohms}$$

The percentage error compared to the exact solution for $\check{z}_c$

$$\text{percentage error} = \frac{149.1 - 127.22}{149.1} \times 100 = 14.68\% \text{ on the low side}$$

C. Computer Estimate of Variance and Distribution

When an equation has several variables and the distributions of each of these variables can be determined, it is possible to use the distribution of the variables to computer generate and graphically determine what form the equation takes. This would also give an independent check on validity for Eqs. (2.16), (2.25), and (2.26) where no previous experience is available. Further, the computer generated data could be used in a solution sizing parts in a coupling equation Eq. (2.42).

V. SAFETY FACTORS AND PROBABILITY OF FAILURE

The applied load $f(a)$ is held in equilibrium by a resisting capacity $f(A)$ of which both will have a distribution due to the variables not being considered as constant values. The desired condition is that the capacity is always greater than the load and the overlap coupling of the two distributions Fig. 2.6 is a small failure value. These should be prescribed values set by the design criterion. The failure values can be found by computer analysis for distributions other than Gaussian or normal functions. However, when

$f(a)$ and $f(A)$ are Gaussian or normal functions [2.18]

$$f(a) = \frac{1}{\breve{z}_a \sqrt{2\pi}} \exp\left[-\frac{1}{2}\left(\frac{a - \mu_a}{\breve{z}_a}\right)^2\right] \tag{2.32}$$

$$f(A) = \frac{1}{\breve{z}_A \sqrt{2\pi}} \exp\left[-\frac{1}{2}\left(\frac{A - \mu_A}{\breve{z}_A}\right)^2\right] \tag{2.33}$$

The range for these evaluations is from minus to plus infinity for a and A. The reliability or probability that capacity is greater than the load is shown in the following equation

$$A - a > 0 \tag{2.34}$$

and letting

$$\zeta = A - a \tag{2.35}$$

then from Eq. (2.17)

$$\mu_\zeta = \mu_A - \mu_a \tag{2.36}$$

$$\breve{z}_\zeta = [\breve{z}_A^2 + \breve{z}_a^2]^{1/2} \tag{2.37}$$

Then $f(\zeta) = f(A) - f(a)$ is a normal distribution which can also be verified by computer with two normal distribution inputs. Then

$$f(\zeta) = \frac{1}{\breve{z}_\zeta \sqrt{2\pi}} \exp\left[-\frac{1}{2}\left(\frac{\zeta - \mu_\zeta}{\breve{z}_\zeta}\right)^2\right] \tag{2.38}$$

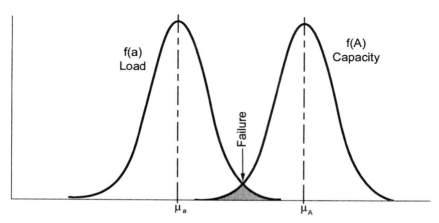

Figure 2.6 Distribution of capacity and load with resulting failure.

Again with a range from minus infinity to plus infinity. The probability of Eq. (2.34) being valid or the reliability $\zeta > 0$ is Eq. (2.38) integrated from zero to   infinity

$$R(\zeta) = \frac{1}{\breve{z}_\zeta \sqrt{2\pi}} \int_0^\infty \exp\left[-\frac{1}{2}\left(\frac{\zeta - \mu_\zeta}{\breve{z}_\zeta}\right)^2\right] d\zeta \tag{2.39}$$

which is the integration of a normal or Gaussian distribution. Using a math handbook for evaluation, let

$$t = \frac{\zeta - \mu_\zeta}{\breve{z}_\zeta} \tag{2.40}$$

when $\zeta = $ zero

$$t = \frac{0 - \mu_\zeta}{\breve{z}_\zeta} = -\frac{\mu_\zeta}{\breve{z}_\zeta}$$

when $\zeta = \infty$

$$t = \frac{\infty - \mu_\zeta}{\breve{z}_\zeta} = \text{infinity}$$

then from a math handbook

$$R(\zeta) = R(t) = \frac{1}{\sqrt{2\pi}} \int_{\frac{\mu_\zeta}{\breve{z}_\zeta}}^\infty \exp\left[-\frac{t^2}{2}\right] dt \tag{2.41}$$

Now the coupling equation is

$$t = -\frac{\mu_\zeta}{\breve{z}_\zeta} = -\frac{\mu_A - \mu_a}{[(\breve{z}_A)^2 + (\breve{z}_a)^2]^{1/2}} \tag{2.42}$$

The $R(t)$ from zero to infinity is 0.5 and from 0 to $-t$ the value added after say $t = 3.5$ add 0.4998 to 0.5 or $R(t) = 0.998$ or

$$R(t) + P(t) = 1 \tag{2.43}$$

Then

$$P(t) = \frac{2}{10^4}$$

which is not accurate enough for a failure rate of one per 10^6 items or more. Table 2.3 shows the value of minus t and the $P(t)$ for more accurate calculations using Eq. (2.42).

Table 2.3 Values of minus t and $P(t)$ for Eqs.
(2.42) and (2.43) with $P(t) = 10^{-D}$

$-t$	D	$-t$	D
zero	infinity	7.3488	13
1.2816	1	7.6506	14
2.3263	2	7.9413	15
3.0912	3	8.2221	16
3.7190	4	8.4938	17
4.2649	5	8.7573	18
4.7534	6	9.0133	19
5.1993	7	9.2623	20
5.6120	8	9.5050	21
5.9478	9	9.7418	22
6.3613	10	9.9730	23
6.7060	11	10.1992	24
7.0345	12	10.4205	25

EXAMPLE 2.13. A material part has a yield coefficient of variation $C_A = \pm 0.07$ and a yield strength mean μ_A of 35,000 psi with an applied mean stress of 20,000 psi, μ_a, and a coefficient of variation of $C_a = \pm 0.10$. Find t for Eq. (2.42) and the reliability and failure.

$$t = -\frac{\mu_A - \mu_a}{[(\breve{z}_A)^2 + (\breve{z}_a)^2]^{1/2}}$$

$$\breve{z}_A = C_A\mu_A = \pm 0.07(35,000 \text{ psi}) \qquad \breve{z}_a = C_a\mu_a = \pm 0.10(20,000 \text{ psi})$$
$$\breve{z}_A = \pm 2450 \text{ psi} \qquad \breve{z}_a = \pm 2000 \text{ psi}$$

$$t = -\frac{35,000 - 20,000}{[(2450)^2 + (2000)^2]^{1/2}} = -4.7428$$

from Table 2.3

$$t = -4.7534 \text{ is } P(t) \approx \frac{1}{10^6} \text{ making } R(t)$$

A value 0.999999. Also note the factor of safety is

$$F.S. = \frac{\mu_A}{\mu_a} = \frac{35}{20} = 1.75$$

Now both $P(t)$ and factor of safety defines the parts safety.

EXAMPLE 2.14. A simple example to give a feel for what can be done with these concepts [2.19]. A tension sample Fig. 2.7 has the following requirements.

$$\text{Load} = (\bar{P}, \check{z}_p) = (6000, 90) \text{ lb}$$

Tensile ultimate 4130 steel = $\bar{F}, \check{z}_F = (156,000; 4300)$ psi

$$P_{failure} = \frac{1}{1000} \qquad R = 0.999 \text{ so } t = -3.0912(\approx -3)$$

The cross-sectional area $A = \pi r^2$
The standard deviation $\check{z}_A = (\partial A / \partial r)dr = 2\pi \bar{r} \check{z}_r$
We are given from manufacturing

$$\check{z}_r = \pm \frac{0.015}{2.576} \; \bar{r} \text{ for 99\% of the samples } z_r = \pm 2.576$$
$$\check{z}_r = 5.83 \times 10^{-3} \; \bar{r} \approx 0.005\bar{r}$$

The applied stress is

$$(\bar{\sigma}, \check{z}_\sigma) = \frac{(\bar{P}, \check{z}_p)}{(\bar{A}, s_A)} = \frac{(6000, 90)}{(\pi \bar{r}^2, 2\pi \bar{r} \check{z}_r)}$$
$$\bar{\sigma} = \frac{\bar{P}}{\pi \bar{r}^2} = \frac{6000}{\pi \bar{r}^2}$$

from Eq. (2.22)

$$\check{z}_\sigma^2 \approx \left[\frac{\bar{A}^2 \check{z}_p^2 + \bar{P}^2 \check{z}_A^2}{\bar{A}^4} \right]$$

with

$$\bar{A} = \pi \bar{r}^2 \quad \bar{P} = 6000 \, \text{lb} \quad \check{z}_A = 2\pi(0.005\bar{r}^2) \quad \check{z}_p = 90 \, \text{lb}$$
$$\check{z}_\sigma^2 \approx \left[1000 \left[\frac{8.1}{\pi^2 \bar{r}^4} + \frac{3.6}{\pi^2 \bar{r}^4} \right] \right] \approx \frac{11,700}{\pi^2 \bar{r}^4}$$

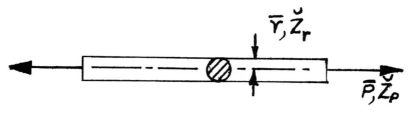

Figure 2.7 A tension sample.

The coupling, Eq. (2.42), is used

$$t = -\frac{\bar{F} - \bar{\sigma}}{\left[\check{z}_F^2 + \check{z}_\sigma^2\right]^{1/2}} = -3$$

with

$$\bar{F} = 156,000\,\text{psi} \qquad \bar{\sigma} = \frac{6000}{\pi\bar{r}^2}$$

$$Z_F = 4300\,\text{psi} \qquad \check{z}_\sigma^2 = \frac{11,700}{\pi^2\bar{r}^4}$$

Substituting and squaring both sides, two solutions for $\bar{r}$ are found. They are $t = -3$ is a structural solution and $t = +3$ for a safety device which is designed to be failed under these conditions.

Structural Member	Safety Device
$R = 0.999 \quad P_f = 0.001$	$R = 0.001 \quad P_f = 0.999$
$\bar{r}_2 = 0.116'' \pm 0.00058''$	$\bar{r}_1 = 0.1055'' \pm 0.00053''$
$\bar{F} = 156,000 \quad \check{z}_F = 4,300$ psi	$\bar{F} = 156,000$ psi $\quad \check{z}_F = 4,300$ psi
$\bar{\sigma}_2 = 141,000$ psi $\quad \check{z}_\sigma = 2,559$ psi	$\bar{\sigma}_1 = 171,500$ psi $\quad \check{z}_\sigma = 3,093$ psi
Safety factor $= \dfrac{156,000}{141,000} = 1.106$	Safety factor $= \dfrac{156,000}{171,500} = 0.909$

The curves are shown in Fig. 2.8 and Fig. 2.9.

EXAMPLE 2.15. Another application of the card sort may be used to develop the standard deviation for the stress due to applied loads.

$$\bar{\sigma} = \frac{P}{A} = \frac{P}{\pi r^2}$$

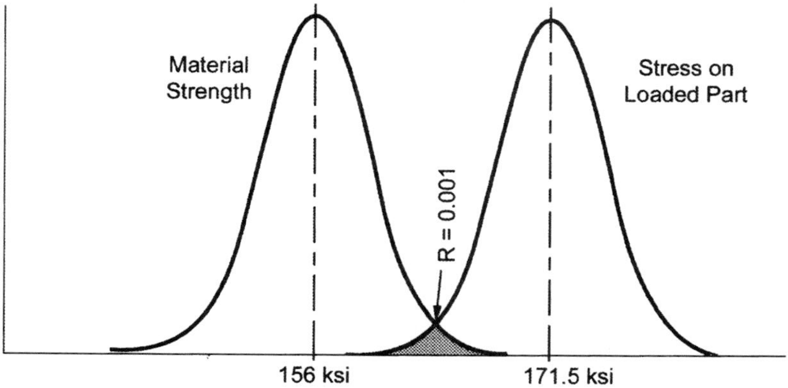

Figure 2.8 Safety device $t = +3$ and $R = 0.001$.

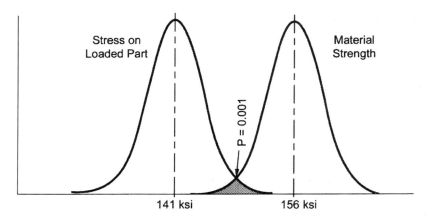

Figure 2.9 Structural member $t = -3$ and $R = 0.999$.

From the Example 2.14 and Eqs. (2.25) and (2.26)

$$r_{max} = 3(0.005\bar{r}) + \bar{r} = 1.015\,\bar{r}$$
$$P_{max} = 6000\,\text{lb} + 3(90\,\text{lb}) = 6270\,\text{lb}$$
$$P_{min} = 6000\,\text{lb} - 3(90\,\text{lb}) = 5730\,\text{lb} \qquad r_{min} = -3(0.005\bar{r}) + \bar{r} = 0.985\,\bar{r}$$

Notes:

1. If P is P_{max} or P_{min} it also applies in both numerator and denominator. In other words one cannot use P_{max} in the numerator and P_{min} in the denominator.
2. The same can be said for r_{max} and r_{min} as for P_{max} and P_{min}.
3. However since the calculation is to find σ_{max} and σ_{min} the following are valid statements

$$\sigma_{max} = \frac{P_{max}}{\pi r_{min}^2}$$

$$\sigma_{min} = \frac{P_{min}}{\pi r_{max}^2}$$

Substituting

$$\sigma_{max} = \frac{6270\,\text{lb}}{\pi[0.985\bar{r}]^2} = \frac{2057.05}{\bar{r}^2}$$

$$\sigma_{min} = \frac{5730\,\text{lb}}{\pi[0.985\bar{r}]^2} = \frac{1770.41}{\bar{r}^2}$$

Again as in the resistor Example 2.10 two variables are selected to obtain σ_{max} and σ_{min} and each is separated from the respective mean by 4.056 standard deviations, 4.056 $\check{z}$

$$2(4.056 \, \check{z}_\sigma) = \sigma_{max} - \sigma_{min}$$

$$\check{z}_\sigma = \left(\frac{2057.05}{\bar{r}^2} - \frac{1770.41}{\bar{r}^2} \right) \frac{1}{2(4.056)}$$

$$\check{z}_\sigma = \frac{35.34}{\bar{r}^2}$$

The previous calculated value, Example 2.14

$$\check{z}_\sigma^2 = \frac{11,700}{\pi^2 \bar{r}^4}$$

$$\check{z}_\sigma = \frac{34.43}{\bar{r}^2}$$

The percent error is the difference of $\check{z}_\sigma$ in Example 2.14 and Example 2.15 divided by $\check{z}_\sigma$ in Example 2.14

$$\% \text{ error} = \frac{(34.43 - 35.34)}{\bar{r}^2 \dfrac{34.43}{\bar{r}^2}} 100$$

$$\% \text{ error} = 2.64\% \text{ on the high side.}$$

Now compare $\check{z}_\sigma = \dfrac{35.34}{\bar{r}^2}$ solution in Eq. (2.42) for Example 2.14

$$t = -3 = \frac{\bar{F} - \bar{\sigma}}{[\check{z}_F^2 + \check{z}_\sigma^2]^{1/2}}$$

Noting

$$\frac{\check{z}_\sigma}{\bar{\sigma}} = \frac{35.34}{\bar{r}^2} \frac{1}{\dfrac{6000}{\pi \bar{r}^2}} = \pm 0.0185$$

Substituting

$$-3 = [156,000 - \sigma][(4300)^2 + (0.0185\sigma)^2]^{-1/2}$$

Squaring and transposing

$$9[(4300)^2 + 3.4240 \times 10^{-4}\sigma^2] = (156,000)^2 - 2(156,000)\sigma + \sigma^2$$

$$[1 - 3.08158 \times 10^{-3}]\sigma^2 - 2(156,000)\sigma - 9(4300)^2 + (156,000)^2 = 0$$

$$A\sigma^2 + B\sigma + C = 0$$

$$\sigma = \frac{(-B) \pm [B^2 - 4A\ C]^{1/2}}{2A}$$

$$\sigma = \frac{2(156,000) \pm [[2(156,000)]^2 - 4(0.99692)[-9(4300)^2 + (156,000)^2]]^{1/2}}{2[0.99692]}$$

$$\sigma = \frac{312,000 \pm 31,039}{2[0.99692]}$$

$\sigma_1 = 172,049$ psi   As before this is a safety device

$\sigma_2 = 140,915$ psi   This is a structural member

$$\sigma_2 = \frac{6000}{\pi \bar{r}^2}$$

$$\bar{r} = \left[\frac{6000}{\pi(140,915)}\right]^{1/2} = 0.1164'' \text{ compared to 0.116 in Example 2.14.}$$

$$\breve{z}_r = 0.005(0.1164) = 0.0006$$

EXAMPLE 2.16. The card sort and partial derivative can be compared to obtain the standard deviation for loading of a cantilever beam for and its stress in Fig. 2.10.

$$\sigma = \frac{MC}{I}$$

$$\sigma = \frac{(PL)h/2}{bh^3/12}$$

$$\sigma = \frac{6PL}{bh^2} = \frac{M}{Z}$$

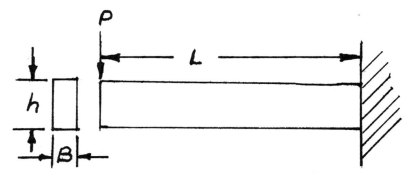

Figure 2.10   Tip loaded cantilever beam.

If

$$C_{v\bar{p}} = \frac{\check{z}_p}{\bar{p}} = \pm 0.01 \quad C_{vb} = \frac{\check{z}_L}{\bar{L}} = \pm 0.01$$

$$C_{vb} = \frac{\check{z}_b}{\bar{b}} = \pm 0.01 \quad C_{vn} = \frac{\check{z}_h}{h} = \pm 0.01$$

for 3 standard deviations

$$P_{\max} = 1.03\bar{p} \quad L_{\max} = 1.03\bar{L} \quad b_{\max} = 1.03\bar{b} \quad h_{\max} = 1.03\bar{h}$$

$$P_{\min} = 0.97\bar{p} \quad L_{\min} = 0.97\bar{L} \quad b_{\min} = 0.97\bar{b} \quad h_{\min} = 0.97\bar{h}$$

$$\sigma_{\max} = \frac{6 p_{\max} L_{\max}}{b_{\min}(h_{\min})^2} = \frac{6(1.03\bar{p})(1.03\bar{L})}{(0.97\bar{b})(0.97\bar{h})^2} = 1.1624\left[\frac{6\bar{p}\bar{L}}{\bar{b}\bar{h}^2}\right]$$

for a card sort, 4 terms selected from Fig. 2.4 and Table 2.2 the spread $\sigma_{\max} - \check{z}_r$ and $\check{z}_r$ is

$$6.0737\,\check{z}_\sigma = \sigma_{\max} = \bar{\sigma}$$

$$\check{z}_\sigma = \frac{1.1624\bar{\sigma} - \bar{\sigma}}{6.0737} = 0.02674\,\bar{\sigma}$$

using the partial derivative method Eq. (2.16)

$$\check{z}_\sigma = \left\{\left[\left(\frac{\partial\sigma}{\partial x_i}\check{z}_{xi}\right)^2\right]\right\}^{1/2}$$

$$\frac{\partial\sigma}{\partial p} = \frac{6\bar{L}}{\bar{b}\bar{h}^2} \quad \frac{\partial\sigma}{\partial L} = \frac{6\bar{p}}{\bar{b}\bar{h}^2}$$

$$\frac{\partial\sigma}{\partial b} = \frac{6\bar{p}\bar{L}}{\bar{h}^2}\left(-\frac{1}{\bar{b}^2}\right) \quad \frac{\partial\sigma}{\partial h} = \frac{6\bar{p}\bar{L}}{\bar{b}^1}\left(-\frac{2\bar{h}}{\bar{h}^4}\right)$$

$$\check{z}_p = \pm 0.01\bar{p} \quad \check{z}_L = \pm 0.01\bar{L} \quad \check{z}_b = \pm 0.01\bar{b} \quad \check{z}_h = \pm 0.01\bar{h}$$

substituting and collecting terms

$$\check{z}_\sigma = \frac{6\bar{p}\bar{L}}{\bar{b}\bar{h}^2}[(0.01)^2 + (0.01)^2 + (0.01)^2 + (2 \times 0.01)^2]^{1/2}$$

$$\check{z}_\sigma = 0.02646\,\bar{\sigma}$$

$$\%\text{error} = \frac{0.02646\bar{\sigma} - 0.02674\bar{\sigma}}{0.02646\bar{\sigma}} \times 100 = 1.06\% \text{ to the high side}$$

for the partial derivative

$$\check{z}_\sigma = \bar{\sigma}[C_{vp}^2 + C_{vL}^2 + C_{vb}^2 + (2C_{vh})^2]^{1/2}$$

VI. FATIGUE

This section uses materials from [2.10] Faupel and Fisher, *Engineering Design*, 2nd Edn (1981) John Wiley and Son Inc. the pages 766–782 and 795–798 are used with the permission of Wiley Liss Inc., a subsidiary of Wiley and Sons Inc. Revisions and additions have been made to reflect the uses of probability.

The material is developed to reflect the probability variations in all of the parameters and to use the concepts in Section V. Authors such as [2.9,2.17,2.26] and others cited are drawn upon to attempt to apply probability to a semi-empirical approach to fatigue through the use of $\sigma_r - \sigma_m$ curves and data concerning the variation of parameters.

The critical loading of a part is in tension under varying loads and temperatures. When the materials are below their high temperature creep limits and above the cold transition temperatures for ductility and operating with a linear stress-strain motion or a reversible one the $\sigma_r - \sigma_m$ curves can be used. The creep limits and cold transition temperatures should be determined for a proposed material as the character will define the thermal limits of a part. Conversely thermal maximums and minimums of a design will define the only materials which can meet the design requirements. The temperatures below the cold transition can be analyzed with $\sigma_r - \sigma_m$ curves with proper corrections for temperature. The problem of elastic buckling may also be considered for the proper fatigue life.

The equations for the fatigue curves are

Soderberg's law $\qquad \dfrac{\sigma_m}{\sigma_y} + \dfrac{\sigma_r}{\sigma_e} = 1$ $\qquad\qquad\qquad\qquad$ (2.44)

Goodman's law $\qquad \dfrac{\sigma_m}{\sigma_u} + \dfrac{\sigma_r}{\sigma_e} = 1$ $\qquad\qquad\qquad\qquad$ (2.45)

Gerber's law $\qquad \left(\dfrac{\sigma_m}{\sigma_u}\right) + \dfrac{\sigma_r}{\sigma_e} = 1$ $\qquad\qquad\qquad\qquad$ (2.46)

σ_r and σ_m are derived from the loading, the part shape and dimensions. The unknown values can be solved for but Eqs. (2.44)–(2.46) will allow only one unknown in each equation. Two or more unknowns require as many equations or an iteration procedure.

If the Soderberg curve, Eq. (2.44), for a simple stress is examined [2.9]

$$\frac{K_1\sigma_m}{\sigma_y} + \frac{K_2\sigma_r}{\sigma_e} = \frac{\sigma_m}{\sigma_y/K_1} + \frac{\sigma_r}{\sigma_e/K_2} = 1 \qquad\qquad (2.47)$$

For all three equations (Eqs. (2.44)–(2.46)), the K_i factors influencing fatigue can be applied either to σ_m and σ_r or σ_y and σ_e. When stresses are complex σ_m and σ_r can be treated using combined stresses, where for plane stress the distortion energy gives

$$\sigma'_m = \sqrt{\sigma_{xm}^2 - \sigma_{xm}\sigma_{ym} + \sigma_{ym}^2 + 3\tau_{xym}^2} \qquad (2.48)$$

$$\sigma'_r = \sqrt{\sigma_{xr}^2 - \sigma_{xr}\sigma_{yr} + \sigma_{yr}^2 + 3\tau_{xyr}^2} \qquad (2.49)$$

The ratio σ'_r/σ'_m and the slope of a line drawn on a $\sigma_r-\sigma_m$ curve from $\sigma_r = \sigma_m = 0$ to intersect the material property line as shown Fig. 2.11. The factor of safety based on the deterministic or average values of loads and dimensions can be determined, however, the probability of failure, p_f, is still not known. The $\sigma_r-\sigma_m$ plots also show R values of stress ratios for slopes and from both the factor of safety is

$$N = A/B \qquad (2.50)$$

the stress variations are related Fig. 2.12 and we see that the alternating component is in each instance that stress which when added to (or subtracted from) the mean stress σ_m the stress variations are related Fig. 2.12 and we see that the alternating component is in each instance that stress which when added to (or subtracted from) the mean stress σ_m results in the

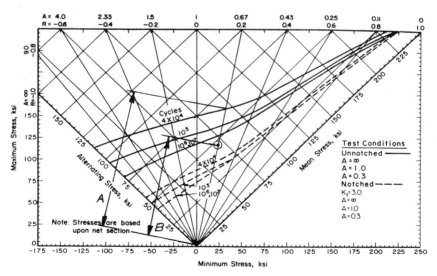

Figure 2.11 Typical constant life fatigue diagram for heat-treated Aisi 4340 alloy steel bar, $F_{tu} = 260$ Ksi [2.65].

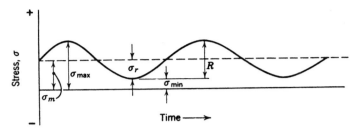

Figure 2.12 Typical sinusoidal fatigue loading with a mean stress.

maximum (or minimum) stress. The average or mean stress σ_m and the alternating component σ_r are Fig. 2.12.

$$\sigma_m = \frac{\sigma_{max} + \sigma_{min}}{2} \quad \text{and} \quad \sigma_r = \frac{\sigma_{max} - \sigma_{min}}{2} \tag{2.51}$$

where a compressive stress is a negative number. For a complete reversal, $\sigma_m = 0$; that is,

$$\sigma_{min} = -\sigma_{max} \quad \text{and} \quad \sigma_r = \sigma_{max}. \text{ In every case,}$$

$$\sigma_{max} = \sigma_m + \sigma_r \tag{2.52}$$

A parameter used to locate the curves of Fig. 2.11 is a stress ratio R defined as

$$R = \frac{\sigma_{min}}{\sigma_{max}}, = \frac{\sigma_m - \sigma_r}{\sigma_m + \sigma_r} \tag{2.53}$$

with stresses used algebraically; $R = -1$ for completely reversed stress, Fig. 2.11.

The curve Fig. 2.11 represents an average for σ_e, σ_u, σ_r, and σ_{ut}. Hence when $N = 1$ the $P_f = 50\%$ which should be avoided. The $\sigma_r - \sigma_m$ curve from extensive testing as per [2.27] will show with the average and spread about the average or mean. The unfortunate case is that only σ_y, σ_u, and σ_e' are generally known as estimates of C_V from much test data. The C_V can be derived from class A and B materials in [2.1,2.18,2.63,2.65] for metallic materials.

$$C_{v\sigma_e} = \pm 0.08 = \frac{\check{z}_{\sigma_{e'}}}{\bar{\sigma}_{e'}}$$

$$C_{v\sigma_{yt}} = \pm 0.07 = \frac{\check{z}_{\sigma_{yt}}}{\bar{\sigma}_{yt}} \tag{2.54}$$

$$C_{v\sigma_{ut}} = \pm 0.05 = \frac{\check{z}_{\sigma_{ut}}}{\bar{\sigma}_{ut}}$$

In order to generate a design curve, σ_e is one of the important factors formulated by Marin and presented by Shigley [2.51] where

$$\sigma_e = k_a k_b k_c k_d k_e k_f \ldots k_1 k_m \sigma_{e'}^1 \tag{2.55}$$

σ_e^1 represents data from a smooth polished rotating beam specimen. The k values can be applied to the stresses or to correct σ_e. Material data can have some k values incorporated in the test or no k values at all. When developing a design curve for combined stresses, it is better to place the k values with the individual stresses where possible. The factors, k values, influencing fatigue behavior will be discussed where most of the corrections are to σ_e or σ_r. σ_e^1 will be discussed in Section VI. B.

A. Some Factors Influencing Fatigue Behavior

The number of variables and combinations of variables that have an influence of the fatigue behavior of parts and structures is discouragingly large, and a thorough discussion concerning this subject is virtually impossible. At best, the designer can make rough estimates and predictions, but even to do this requires some knowledge of at least the various principal factors involved. In the following discussion some high-spot information is presented with the caution that fatigue behavior is extremely complicated and any data or methods of utilizing the data should be viewed in a most critical way.

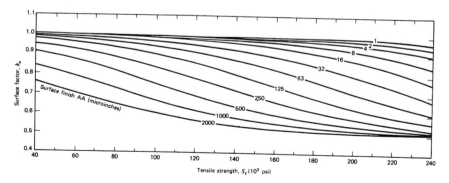

Figure 2.13 k_a versus surface roughness and tensile strength (after Johnson [2.25] courtesy of machine design).

1. Surface Condition, k_a

By surface condition is meant the degree of smoothness of the part and the
presence or absence of corrosive effects. In general, a highly polished surface
gives the highest fatigue life, although there is evidence suggesting that the
uniformity of finish is more important than the finish itself. For example,
a single scratch on a highly polished surface would probably lead to a fatigue
life somewhat lower than for a surface containing an even distribution of
scratches. Typical trend data of Karpov and reported by Landau [2.30]
are shown in Fig. 2.14 for steel. Reference [2.32] also shows data for forgings
that are similar to the k_a for tap water. The *Machinery Handbook* [2.62]
shows a detailed breakdown of surface roughness versus machining or
casting processes. This information can be used for steels to find the k_a from
a theoretical model development by Johnson [2.25] in Fig. 2.13.

The data for k_a is plotted with the equations derived by [2.18] from
data shown in [2.9] for steel.

Ground:

$$k_a = 1.006 - 0.715 \times 10^{-6}\bar{\sigma}_{ult} \tag{2.56}$$

Machined:

$$k_a = 0.947 - 0.159 \times 10^{-5}\bar{\sigma}_{ult} \tag{2.57}$$

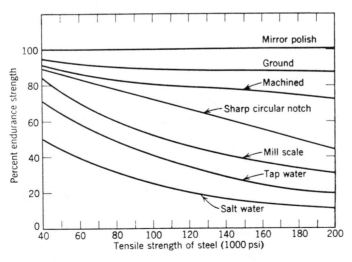

Figure 2.14 Effect of surface condition on fatigue of steel.

Hot rolled:

$$k_a = \frac{20919 + 0.0545\bar{\sigma}_{ult}}{(\bar{\sigma}_{ult}/2)} \qquad \text{a parabolic form} \qquad (2.58)$$

As forged:

$$k_a = \frac{20955 - 0.00266\bar{\sigma}_{ult}}{(\bar{\sigma}_{ult}/2)} \qquad \text{a parabolic form} \qquad (2.59)$$

The standard deviations [2.18] and coefficients of variation [2.51] are in Table 2.4.

2. Size and Shape, k_b

The subject of size and shape effects in design is discussed; the same general conclusions and methods presented here also apply to fatigue loading. For example, it is seen that the small bar has less volume of material exposed to a high stress condition for a given loading and consequently should exhibit a higher fatigue life than the larger bar. Some data illustrating this effect are shown [2.15]. Shape (moment of inertia) also has an effect as shown [2.15]. In design it is important to consider effects of size and shape, but by proper attention to these factors a part several inches in diameter can be designed to on the basis of fatigue data obtained on small specimens. A rough guide presented by Castleberry, Juvinall, and Shigley is

$$k_b = \begin{bmatrix} 1 & \text{for } d \leq 0.30\text{in.}(2.26, 2.51) \\ 0.85 & 0.3 \leq d \leq 2\text{in.}(2.26, 2.51) \\ 1 - \dfrac{(d-0.30)}{15} & 2 \leq d \leq 9\text{in.}(2.4) \\ 0.65 - 0.75d & 4 \leq d \leq 12(2.18) \\ c_v \approx 5 \text{ to } 6\% & \min(2.18) \end{bmatrix} \qquad (2.60)$$

Table 2.4 k_a standard deviations $\check{z}_a$ and coefficients of variation

Surface finish	Shigley, Miscke [2.51]	Haugen [2.18]
Ground	$C_V = 0.13$	$\check{z}_a = 0.103$
Cold drawn & machined	$C_V = 0.06$	$\check{z}_a = 0.0406$
Hot rolled	$C_V = 0.11$	$\check{z}_a = \dfrac{2780.5}{(\bar{\sigma}_{ult}/2)}$
Forged	$C_V = 0.08$	$\check{z}_a = \dfrac{2780.5}{(\bar{\sigma}_{ult}/2)}$

The k_b, greater than 0.5, is for steel and only serves as a guide to other materials. When $d \leq 0.30$ many materials fall into the range of spring diameters where ultimate and endurance limit strengths [2.9] are stated as a function of wire diameter and the k_b is greater than one. This also applys for constant material thickness.

3. Reliability, k_c

The k_c value corrects σ_e^1 for an 8% standard deviation when no other data are available. In Fig. 2.15 the $\sigma_r - \sigma_m$ curve can be developed; however, the solid line is the average or mean of all data. The reliability of a design using any point on the solid line is 0.50. Tests have been conducted [2.27] where the dotted lines A and B represent data spread of $\pm 3\sigma$ derived from several tests along the curve. Eq. (2.61)–(2.64)

    The k_c values [2.51] are as follows:

$$R(0.50) = 1.00 \tag{2.61}$$

$$R(0.90) = 0.897 \tag{2.62}$$

$$R(0.95) = 0.868 \tag{2.63}$$

$$R(0.99) = 0.814 \tag{2.64}$$

Note: this correction is used when attempting to derive a design line to be

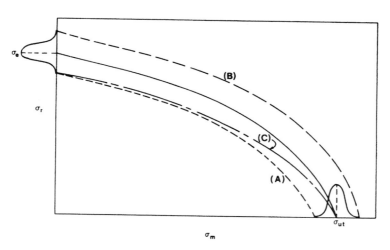

Figure 2.15 $\sigma_r - \sigma_m$ with material property variations.

used with a factor of safety calculation. Further note (Eq. (2.54))

$$C_v = \pm 0.08 \text{ for } \sigma_e^1$$
$$C_v = \pm 0.07 \text{ for metallic yields}$$
$$C_v = \pm 0.05 \text{ for metallic ultimates}$$

$k_c = 1$ for a coupling equation calculation (Eq. (2.42)) and $\check{z}_c = 0$.

The values correct σ_e or increase the amplitude stress depending on the calculation. If a curve is developed and line A in Fig. 2.15 is to be drawn, some knowledge about the spread of the data about the σ_{ut} mean should be obtained. However, if k_c is only used on σ_e^1, it should be noted that a line c is generated where the reliability is 0.99 at σ_e and slips to 0.50 at the ultimate, σ_{ut}. The curve would be more accurate if k_c is applied to both σ_{ut} and σ_e until material data are available.

4. Temperature, k_d

In general, the endurance limit increases as temperature decreases, but specific data should be obtained for any anticipated temperature condition since factors other than temperature, per se, could control. For example, for many steels the range of temperature associated with transition from ductile to brittle behavior has to be allowed for. In addition, for some materials, structural phase changes occur at elevated temperature that might tend to increase the fatigue life. The low temperature k_d values [2.11] for $-186°C$ to $-196°C$ are approximately in Table 2.5.

The values decrease linearly with temperature to the room temperature value of one. These k_d values increase σ_e and decrease σ_m and σ_r.

Unlike low temperature values, k_d is not linear above room temperatures for metals. Typical k_d values are in Table 2.6.

Many k_d values for specific alloys and temperatures can be found in [2.1,2.11,2.65]. The actual $\sigma_r - \sigma_m$ curves are available for many materials at elevated and cyrogenic temperatures with σ_e^1 and σ_{ult} test values. The

Table 2.5 Low temperature correction, k_d, for metals

Carbon steels	2.57
Alloy steels	1.61
Stainless steels	1.54
Aluminum alloys	1.14
Titanium alloys	1.40

Table 2.6 Metal correction values, k_d, above room temperature

Magnesium	(572°F)	0.4
Aluminum	(662°F)	0.24
Cast alloys	(500°F)	0.55
Titanium	(752°F)	0.70
Heat resistant steel	(1382°F)	0.63
Nickel alloys	(1382°F)	0.70

curves are 50 percentile curves and the variations on σ_e^1 σ_{yt}, and σ_{ult} still must be estimated with known C_v data.

5. Stress Concentration, k_e

The subject of stress concentration is considered separately in [2.10,2.49] and the discussion concerning the effect of mechanical stress concentrators such as grooves, notches, and so on, on fatigue behavior is included as part of [2.10,2.49]. Later in this chapter the effect of stress concentrations such as inclusions in the material is considered. In general, the presence of any kind of a stress raiser lowers the fatigue life of a part or structure.

Stress concentrations are introduced in two ways:

a. The geometry of a design and loading creates stress concentrations Fig. 2.16. This is introduced into the design calculations by

$$q = \frac{K_f - 1}{K_t - 1} \tag{2.65}$$

where q = the notch sensitivity factor [2.9,2.51,2.54]
$\quad K_t$ = theoretical factors [2.9,2.51,2.54]
Often to find q a notch radius, r, is required which is generally not known until the design is completed. Therefore, to start a design

$$K_f = K_t \tag{2.66}$$

K_f is used in Eq. (2.55) to correct σ_e for a single state of stress as

$$k_e = \frac{1}{K_f} \tag{2.67}$$

Otherwise, for combined states of stress K_f is used in Eq. (2.49) for the design of ductile materials and in Eqs. (2.48) and (2.49) for the design of brittle materials. For example: for a ductile material with a bending stress

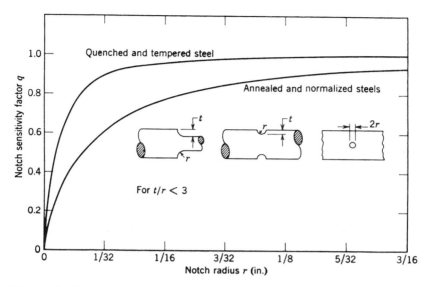

Figure 2.16 Typical notch sensitivity data for steel (data by Peterson [2.49]).

and an axial load.

$$\sigma_{xr} = K_{fb}\frac{M_a c}{I} + K_{ft}\frac{P_a}{A} \tag{2.68}$$

where K_{fb} and K_{ft} stand for the stress concentration factors for bending and tension.

In general K_f is from Eq. (2.65) where

$$K_f = 1 + q(K_t - 1)$$

If $q = 1$ then $K_f = K_t$ which is the first iteration of a part size then K_f is calculated when the notch radii and part across sections dimensions are known. The variations are in K_t and q. Haugen [2.18] has plotted and calculated K_t for some shapes. The C_vs are $\approx 10.9\%$ with $R(0.99)$ and 95% confidence for the smaller radii and higher K_ts but the 10.9% becomes smaller as the K_t curve flattens out.

The q average values are published in most texts but in [2.52] and [2.36] the coefficient of variation for q, and C_v, Table 2.7, may be developed and the estimates are as follows

EXAMPLE 2.17. The card sort may be used to find $\check{z}_f$ for the maximum or minimum variables where a symmetric distribution is used, such as tolerance on a part size

Table 2.7 C_v for q average values

	Q and T steel	Normal steel	Ave. aluminum
C_v	±8.33%	±5.26%	±7.33%

In the equation

$$K_f = 1 + q(K_t - 1)$$

For q_{high}

$$q_{high} = \bar{q}(1 + 3C_{vq}) = 1.2499\ \bar{q} = \bar{q}[+3(0.0833)]$$
$$K_{thigh} = \bar{K}_t(1 + 3Cvkt) = 1.327\ \bar{K}_t = \bar{K}_t[1 + 3(0.109)]$$

If $\bar{q} = 1$ and $K_t = 2.6$

$$\bar{K}_f = 1 + (1)(2.6 - 1) = 2.6$$

$$K_{f_{high}} = 1 + [1.2499(1)][1.327(2.6) - 1] = 4.0625$$
Now

$$4.0556\ \check{z}_f = K_{f_{high}} - \bar{K}_f$$

The 4.0556 is from Fig. 2.4 and Table 2.2 where $P(q_{high}K_{thigh})$ acts as 4.0556 standard deviations.

$$\check{z}_f = \frac{4.0625 - 2.6}{4.0556} = 0.3606$$

The

$$C_V = \frac{\check{z}_f}{\bar{K}_f} = \frac{0.3606}{2.6} \times 100 = \pm 13.87\%$$

Should the sort be taken as a one sided distribution Fig. 2.17
Fig. 2.17 one sided distribution K_f

$$q_{high} = \bar{q}(1 + 2.5758Cvq) = 1.2146\ \bar{q}$$
$$K_{thigh} = \bar{K}_t(1 + 2.5758Cvkt) = 1.2808\ \bar{K}_t$$

Substituting

$$K_{f_{high}} = 1 + [1.2146(1)][1.2808(2.6) - 1] = 3.8301$$

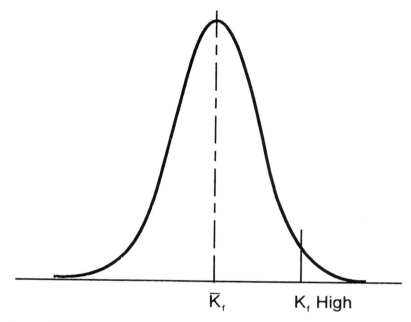

Figure 2.17 One sided distribution for K_f.

again

$$4.0556 \, \breve{z}_f = 3.8301 - 2.6 = K_{fhigh} - \bar{K}_f$$

$$\breve{z}_f = \frac{3.8301 - 2.6}{4.0556} = 0.3033$$

$$C_v = \frac{\breve{z}_f}{\bar{K}_f} = \frac{0.3033}{2.6} \times 100 = \pm 11.66\%$$

(2.69)

Here the choice is 11.66% or 13.87% depending if the sort is taken from a one sided on a two sided distribution. Shigley and Miscke (2.51) quotes 8–13% for C_{vkf} for steel samples of various notch shapes.

  b. Stress concentrations also develop with cracks in a material whether caused by machining, heat treatment, or a flaw in the material. Therefore, some consideration should be given to crack size. In Fig. 2.19 grooves or cracks in polished samples [2.11] for 1–10 rms finish are less than 0.001 mm (3.95×10^{-5} in.) while in a rough turn, 190–1500 rms, the cracks are 0.025–0.050 mm (0.001–0.002 in.) long. The minimum detectable crack [2.34] with X-ray or fluoroscope is about 0.16 mm (0.006 in.). In Fig. 2.18

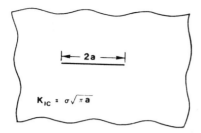

Figure 2.18   Inherent flaw in a large part.

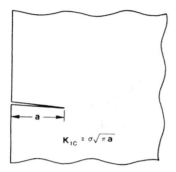

Figure 2.19   Surface crack in a large part.

the inherent flaws [2.12] in steel, 2a length, under the surface run from 0.001 to 0.004 in. decreasing with strength. Aluminum and magnesium alloys vary from 0.003–0.004 in. while copper alloy has, 2a crack length, of up to 0.007 in.

In [2.10] cracks are discussed as well as what K_{IC} and K_{th} mean in fatigue. When cracks grow [2.16] K_{th} is exceeded, and when cracks split a part into pieces K_{IC} has been exceeded.

$$\frac{1}{\pi}\left(1.5\frac{E}{\sigma}\times 10^{-4}\right)^2 \le a_{th} \le \frac{1}{\pi}\left(1.8\frac{E}{\sigma}\times 10^{-4}\right)^2 \tag{2.70}$$

where E is the modulus and σ is the highest stress. Work by Siebel and Gaier presented by Forrest [2.12] on machining grooves will be compared with a_{th} in Table 2.8. An operating stress of 30 kpsi is selected for the illustration. The representative rms values are from [2.30,2.62].

Table 2.8 a_{th} compared to machined grooves

	rms	Groove depth (inches $\times 10^{-3}$)	a_{th} (30 kpsi) (inches $\times 10^{-3}$)
Polish	8	0.04	Steel (7–10)
Fine grind	10	0.08	Aluminum (0.8–1.2)
Rough grind	70	0.2–0.4	Magnesium (0.3–0.5)
Fine turn	10–90	0.4–0.8	Titanium (2–3)
Rough turn	90–500	0.8–2	
Very rough turn	>500	>2	

In (2.1) (2.18) (2.65) the C_{vkic} is

$$1\% \leq C_{vkic} \leq 28\% \tag{2.71}$$

For various materials fabricated by rolling, forging, also the forming directions, and thickness of the samples. Each material must be researched for applicable data and the variation of k_{ic} is not straight forward and easily expressed.

The surface crack in Table 2.8 and Fig. 2.19 is accounted for in k_a, surface conditions, and its effects are further reduced by residual stresses k_f, surface treatment k_h, and discussed in fretting k_i. However, the inherent flaw (Fig. 2.18) must be detected by nondestructive testing such as x-rays. Then, the part is either scrapped or repaired.

6. Residual Stress, k_f

The subject of residual stress is considered separately [2.10] which may be referred to for more details. For present purposes it is to be noted that, in general, a favorable residual stress distribution in a part leads to an increased fatigue life; typical applications are shot peening or surface rolling of shafts and autofrettage of cylinders.

Shot peening on any part surface–whether it be machined, surface hardened, or plated–will generally increase endurance strength. The shot peening residual stress is compressive and generally half of the yield strength and with a depth of 0.020–0.040 in. The shot peening effect [2.9] disappears for steel above 500°F and for aluminum above 250°F. The correction to the endurance strength for shot peening is

$$k_f = (1 + Y) \tag{2.72}$$

where Y is the improvement.

Typical values for steel are shown in Table 2.9

The roughest surface will realize the largest values of Y improvement. However, the overall net effect of shot peening is to increase σ_e so that

$$0.70\sigma_e^1 \leq \sigma_e \leq 0.90\sigma_e^1 \tag{2.73}$$

where σ_e^1 is the endurance strength of a mirror-polished test sample, and σ_e is calculated from

$$\sigma_e = k_a k_f \sigma_e^1 \tag{2.74}$$

Surface rolling induces a deeper layer than shot peening (0.040 to 0.05 in.). The Y improvements are shown in Table 2.10.

Actual cases with discussion are presented by Frost [2.12], Forrest [2.11], Faires [2.9], Lipson and Juvinall [2.32] as well as [2.54,2.63 Vol. II; 2.8].

Cold-working of axles also imparts compressive residual stresses that tend to increase the fatigue life. If, however, collars are press-fitted on shafts, an effective stress raiser is formed at the interface (Fig. 2.20) which offsets any beneficial effect of the compressive residual stress and usually results in a lower fatigue life. This difficulty may be overcome to a large extent by the modified arrangements of the collar shown.

Table 2.9 Shot peening improvements for steel fabrication

Surface	Y	$\bar{Y}$	C_{v_y}
Polished	0.04–0.22	0.13	23%
Machined	0.25		
Rolled	0.25–0.5	0.375	11.1%
Forged	1–2	1.5	11.1%

Table 2.10 Surface rolling improvements for materials

Surface or material	$\bar{Y}$	Y	C_{v_y}
Straight steel shafts	0.5	0.2–0.8	20%
Polished or machined steel parts	0.28	0.06–0.5	26.2%
Magnesium		0.5	
Aluminum	0.25	0.2–0.3	6.7%
Cast iron	1.065	0.2–1.93	27.1%
Any condition	0.50	0.1–0.9	26.7%

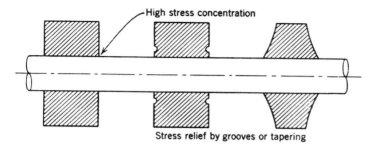

Figure 2.20 Various assemblies of collars shrunk on a shaft.

EXAMPLE. Straight shaft of steel with surface rolling find $C_{v_{k_f}}$
Table 2.10
where

$$\bar{k}_f = 1 + \bar{y} = 1.5$$
$$k_{f min d} = 1 + \bar{y}(1 - 3C_{vy}) = 1 + 0.5(1 - 3[0.2]) = 1.2$$
$$k_{f max} = 1 + \bar{y}(1 + 3C_{vy}) = 1 + 0.5(1 + 3[0.2]) = 1.8$$

Since 1 variable is used as a card sort and the range is six standard deviations

$$6\breve{z}_f = k_{f max} - k_{min}$$
$$\breve{z}_f = \frac{1.8 - 1.2}{6} = \frac{0.6}{6} \tag{2.75}$$
$$C_{v_{k_f}} = \frac{\breve{z}_f}{\bar{k}_f} = \frac{0.1}{1.5} \times 100 = 6.67\%$$

7. Internal Structure, k_g

For the purposes of this book the only internal structural aspects of fatigue behavior of materials of interest are inclusions that act as stress concentrators and (probably related to inclusions) directional effects giving rise to different fatigue properties in the longitudinal and transverse directions of fabricated materials. By longitudinal is meant the axis of rolling direction in sheet, for example. More will be said about this later in the design application examples where it is pointed out that the transverse fatigue properties of many steels, for example, are distinctly lower than the longitudinal fatigue properties. Here much data is required. Generally $kg = 1$, C_v undefined, $\breve{z}_g = 0$.

8. Environment, k_h

The effects of tap and salt water on steel are shown in Figure 2.14. The same effects for nonferrous metal [2.11,2.47] for all tensile strengths over $6 \check{z}_{k_h}$ are

$$0.40 \leq k_h \leq 0.64 \quad k_h(0.52, 0.040) \tag{2.76}$$

Two exceptions are electrolytic copper and copper-nickel alloys for which

$$0.85 \leq k_h \leq 1.06, \quad k_h(0.955, 0.035) \tag{2.77}$$

and nickel-copper alloys for which

$$0.64 \leq k_h \leq 0.86, \quad k_h(0.75, 0.0367) \tag{2.78}$$

These results are from tests conducted from 1930–1950; therefore, care should be taken with newer alloys.
The effect of steam on steel under pressure is

$$0.70 \leq k_h \leq 0.94, \quad k_h(0.82, 0.040) \tag{2.79}$$

However, for a jet of steam acting in air on steel the values are one-half of Eq. (2.79).
A corrosive environment on anodized aluminum and magnesium yields

$$0.76 \leq k_h \leq 1, \quad k_h(0.88, 0.040) \tag{2.80}$$

while for nitrided steel

$$0.68 \leq k_h \leq 0.80 \quad k_h(0.74, \ 0.020) \tag{2.81}$$

9. Surface Treatment and Hardening, k_i

Surface treatment protects the surface from gross corrosion. Results [2.12] for chrome plating of steel are represented by the reduction of the fatigue strength

$$Y = 0.3667 - 9.193 \times 10^{-3}\sigma_e^1 \tag{2.82}$$

where σ_e^1 is the fatigue strength of the base material. Then k_i is

$$k_i = 1 + Y \tag{2.83}$$

For nickel plating [2.54], Y is -0.99 in 1008 steel and -0.33 in 1063 steel. If shot peening [2.9] is performed after nickel and chrome plating, the fatigue strength can be increased above that of the base metal.

The endurance strength of anodized aluminum in general is not affected. Osgood [2.47] and Forrest [2.11] present effects of other surface treatments.

Surface hardening of steel produces a hardened layer to resist wear and cracking. Table 2.11 compares typical layer thicknesses and increases in Y from [2.8,2.9,2.11,2.12,2.63 Vol II]. Rotating beam samples exhibit Y values of 0.20–1.05 for carburizing or nitriding. The soft layer under the surface hardening should always be checked to see if its endurance strength is adequate. In the soft layer the tension residual stresses will balance the compression stresses in the outer layer.

10.   Fretting, k_j

Fretting can occur in parts where motions of 0.0001–0.004 inches maximum take place between two surfaces. The surface pairs exist as

1.   Tapered cone and shaft assemblies
2.   Pin, bolted, or riveted joints
3.   Leaf springs
4.   Ball and bearing race
5.   Mechanical slides under vibration
6.   Spline connections
7.   Spring connections
8.   Keyed shafts and joints

These surfaces, under pressure, work against each other producing pits and metal particles. Extensive action can result in cracks and finally failure.

Table 2.11   Surface hardening–Y increases and layer thickness, inches

	Flame and induction hardening	Carburizing	Nitriding
Layer thickness (inches)	0.125–0.500 (induction) $\approx$0.125 (flame)	0.03–0.1	0.004–0.02
Steel	$(\bar{y}, \check{z}_y) =$ (0.73, 0.0233) 0.66–0.80	(0.735, 0.038)– 0.62–0.85	
Alloy steel	(0.35, 0.0967) 0.06–0.64	(0.19, 0.0567) 0.02–0.36	(0.65, 0.1167) 0.30–1.00
Rotation beam samples		(0.625, 0.1417) 0.2–1.05	(0.625, 0.1417) 0.2–1.05

The action can be recognized in disassembled parts by a rust color residue in ferrous parts and by a black residue for aluminum and magnesium parts. Desirable surface pairs [2.11, 2.26, 2.47] can be selected to reduce fretting. Steel surfaces react well while cast iron must be lubricated to obtain the same performance.

Sors [2.54] concludes k_j is 0.70–0.8 $(\bar{k}_j, \check{z}_{kj}) = (0.75, 0.0167)$ in general and 0.95 for good surface matches. Frost [2.12] reports similar values with some surface pairs lower.

Fretting may be reduced [2.8, 2.9, 2.15] by constraining the motion or by closer fits, lubricated surfaces or gaskets, and residual stresses imparted to the surfaces. A constraining example: a flywheel on a shaft with a keyed cone fit held in place with a loading nut. Surface lubrication with molybdenum disulfide, MoS_2 as well as other inhibitors extends the life of a surface before fretting. Further, reductions can be obtained if the surfaces are shot peened or surface rolled prior to assembly.

11. Shock or Vibration Loading, k_k

These effects increase both σ_r and σ_m and in essence decreases the life of a designed part or system. The most effective method is to develop the stresses from the loading because of the extensive methods involved, such as structural dynamic programs and even models loaded with quasi-static design accelerations. A quasi-static design can be used to estimate a desired fatigue life.

The quasi-static loads are educated guesses at what loads a data based design criterion is telling the designer. The fact that some material properties change with the rate of loading, mostly for the better, should be kept in mind. The damping for quasi-static loads is assumed or guessed (this also happens in computer models). In Eq. (2.55) for σ_e the value of $k_k = 1$ and $\check{z}_k = 0$ for a calculation of $\check{z}_{\sigma_e}$.

12. Radiation, k_l

Radiation tends to increase tensile strength and decrease ductility. The effect is discussed in more detail in [2.47] Unless data is available $k_l = 1$, $C_{v_{k_L}} = 0$.

13. Speed

For most metallic materials and other materials that do not have viscoelastic properties, stress frequencies in the range from about 200–7000 cycles per minute have little or not effect on fatigue life. The fatigue life could be affected, however, if during the rapid stress fluctuations, the temperature increased appreciably. For speeds over 7000 cycles per minute there is some evidence that the fatigue life increases a small amount.

For viscoelastic materials (polymers), considerably more caution must be exercised in interpreting fatigue data. Normally, fatigue tests are conducted at as rapid a frequency of stressing as possible, with due consideration for temperature rise. However, polymers will exhibit different fatigue characteristics depending on the stress frequency, which, depending on the material, will yield different results in ranges of high and low loss factors. In general, in applications involving fatigue loading it is best to use materials that exhibit low loss factor under the conditions expected to persist in the application; if the part is used as a damper or energy absorber, it should be used at a frequency characterized by a high loss factor. For example, a vibrating part made of polymethyl methacrylate at room temperature should not be used at a frequency of 600 cycles per minute since this is where the loss factor is maximum (Fig. 2.21). Thus, in evaluating fatigue data for polymers curves such as shown in Fig. 2.21 should be available.

14. Mean Stress

A structure stress-cycled about some mean stress other than zero has different fatigue characteristics than one cycled about zero mean stress. The precise reason for this is unknown, but it is believed due to hysteresis effects caused by plastic flow that changes the fatigue characteristics on each cycle. The effect of mean stress has been included empirically in the design examples discussed later in this section.

The mean stress, Eq. (2.48), for brittle metals requires the application of K_f values as shown in Eq. (2.68). For some steels, for example, the criterion for brittleness can be found approximately from Charpy or Izod test data shown in Fig. 2.22. Above the transition temperature the metal acts in a ductile manner while below the transition temperature the metal acts in a brittle fashion.

The combined mean stress Fig. 2.23 using distortion energy or Von Mises energy criterion is Eq. (2.48)

$$\sigma_m^1 = \sqrt{\sigma_{xm}^2 - \sigma_{xm}\sigma_{ym} + \sigma_{ym}^2 + 3\tau_{xym}^2} \qquad (2.84)$$

$\breve{z}_{\sigma_m^1}$ the standard deviation needs to be calculated or a distribution function developed.

The reversal Fig. 2.24 or amplitude stress Eq. (2.49)

$$\sigma_r^1 = \sqrt{\sigma_{xa}^2 - \sigma_{xa}\sigma_{ya} + \sigma_{ya}^2 + 3\tau_{xya}^2} \qquad (2.85)$$

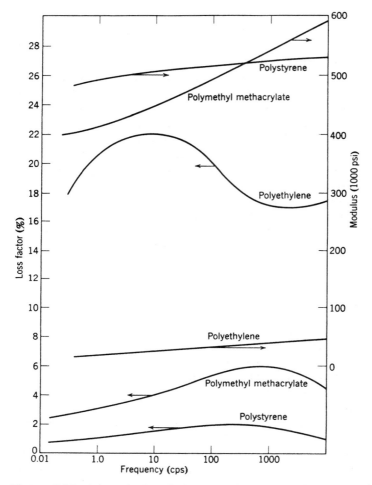

Figure 2.21 Dynamic data for three polymers at room temperature.

Here $\breve{z}_{\sigma r}$, the standard deviation needs development. Further, note, for ductile materials K_f will be applied to the stress condition for σ'_r and also to σ'_m for brittle materials. In calculating σ'_r a simple but important case is a rotating beam subjected to a constant bending moment. For no stress concentration the stress Fig. 2.25 is

$$\sigma_{r'} = \sigma_{xa} = \pm \frac{M_m c}{I} \tag{2.86}$$

for a reversal stress from a constant bending moment.

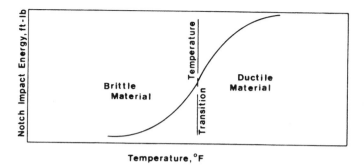

Figure 2.22 Typical charpy or Izod test of steel.

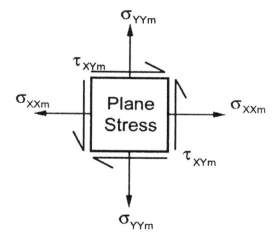

Figure 2.23 Mean plane stress.

When the shaft is not rotating ($\omega = 0$)

$$\sigma_r^1 = \pm \frac{M_a c}{I} \tag{2.87}$$

$$\sigma_m^1 = \pm \frac{M_m c}{I} \tag{2.88}$$

B. Fatigue Properties of Materials

Fatigue life of a material is not a property like modulus, which, under normal conditions, is a material constant. The endurance limit of a material

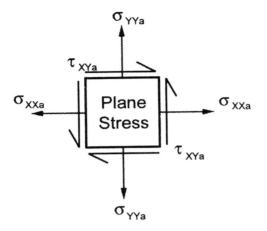

Figure 2.24   Amplitude plane stress.

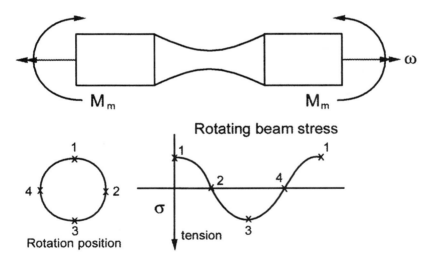

Figure 2.25   A reversal stress from a constant bending moment in a rotating beam.

is influenced by the type of test used and numerous other variables, of which many have already been mentioned. Therefore for any particular material it is necessary to examine fatigue data with respect to the end use conditions. Nevertheless, as a guide, a few properties are presented here to assist in the selection of materials for particular applications. More detailed property information on specific materials should be obtained from the fabricator or vendor. Information can also be obtained from [2.1, 2.63, 2.64, 2.66] as well as from publications like *Machine Design, Modern Plastics, and Materials Engineering*, which publish data yearly. Most organizations that publish standards and mechanical strength properties such as SAE, ASME, AISC, AITC, and others are listed in [2.60].

As previously mentioned most ferrous materials are characterized by a more or less definite endurance limit which is of the order of half the ultimate tensile strength of the material. Typical data for a wide range of ferrous materials are shown in Fig. 2.26. In using such data it is necessary to consider the fact that most steels exhibit anisotropy of fatigue properties and that the values reported in the curves (like Fig. 2.26) are probably from tests of specimens cut in the longitudinal direction. The annealed austenitic stainless steels have very good fatigue-corrosion resistance and are not as notch-sensitive as other steels; however, in the cold-worked condition, their fatigue properties are about the same as those exhibited by other steels. Typical fatigue data for a variety of other materials are shown in Figs. 2.27 to 2.31.

The σ_e^1 data for fatigue strength is presented versus strength [2.15], low cycle fatigue ($<10^3$ cycles) material data for cylic loading [2.1, 2.3] and [2.65]. The design life in cycles may be selected. When designing for more than 10^8 or 10^6 cycles [2.2] a reduction value may be used here called k_m (in [2.2], this is called K_L and C_L) the values are:

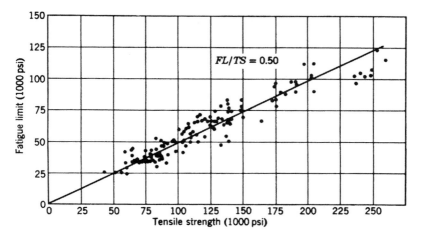

Figure 2.26 Rotating-bending fatigue limits of cast and wrought steels at one-million cycles. (After Grover, Gordon, and Jackson [2.15]. Courtesy of naval weapons, U.S. Department of the Navy). [$\breve{z}_{\sigma_c'} = \sigma_{ULT/30}$; $\sigma_e^1 \leq 100,00$]

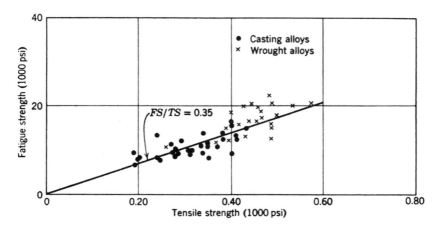

Figure 2.27 Rotating-bending strengths at 100 million cycles of magnesium alloys. (After Grover, Gordon, and Jackson [2.15]. Courtesy of the Bureau of Naval Weapons, U.S. Department of the Navy). $[\breve{z}_{\sigma'_e} = \sigma_{ULT/20}]$

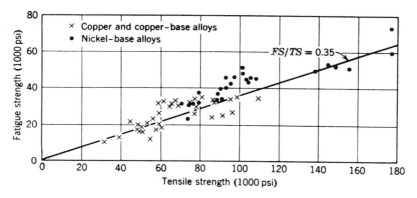

Figure 2.28 Rotating-bending fatigue strengths at 100 million cycles of some nonferrous alloys. (After Grover, Gordon, and Jackson [2.15]. Courtesy of the Bureau of Naval Weapons, U.S. Department of the Navy). $[\breve{z}_{\sigma'_e} = \sigma_{ULT/50}]$

1. *Bending*

$$k_m = (\bar{k}_m, \breve{z}_m) = (0.85, 0.01967)@\,10^{10} \text{ cycles for gear materials} \tag{2.89}$$

$$= 1@\,10^7 \text{ cycles and } \breve{z}_m = 0 \tag{2.90}$$

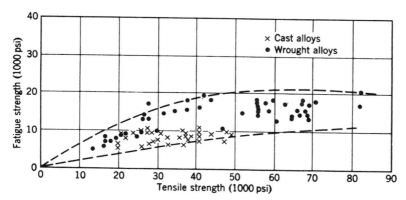

Figure 2.29 Rotating-bending fatigue strengths at 500 millions cycles of aluminum alloys. (After Grover, Gordon, and Jackson [2.15]. Courtesy of the Bureau of Naval Weapons, U.S. Department of the Navy). [Wrought $\breve{z}_{\sigma_e^l} = \frac{500}{3}$; CAST $= \frac{2500}{3}$]

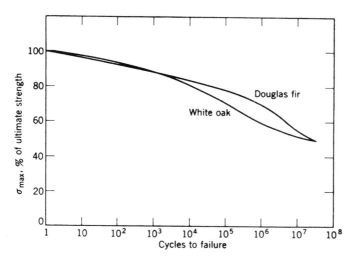

Figure 2.30 Typical fatigue test data for wood. Repeated tension parallel to grain at 900 cycles per min (see [2.40]).

2. *Contact*

$$k_m = (\bar{k}_m, \breve{z}_m) = (0.7625, 0.0383)@10^{10} \text{ cycles for gear materials} \tag{2.91}$$

$$= 1@10^7 \text{ cycles and } \breve{z}_m = 0 \tag{2.92}$$

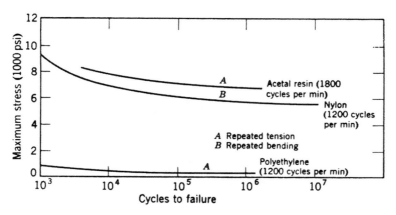

Figure 2.31 Typical fatigue test data for three polymers.

The scatter bands [2.15] show for aluminium alloys at 5×10^8 cycles sand cast:

$$\frac{\check{z}_{\sigma_e^1}}{\sigma_e^1} = \pm 11.11\% \tag{2.93}$$

permanent mold castings:

$$\frac{\check{z}_{\sigma_e^1}}{\sigma_e^1} = \pm 12.31\% \tag{2.94}$$

Wrought alloys:

$$\frac{\check{z}_{\sigma_e^1}}{\sigma_e^1} = \pm 7.07\% \tag{2.95}$$

Titanium alloys [2.11] behave as cast and wrought steel in Fig. 2.26. Wood, polymers with low modulus filler, and plywood [2.11, 2.15, 2.54] for 10^7 cycles have

$$0.20 \leq \frac{FS}{TS} \leq 0.40. \tag{2.96}$$

Unidirectional nonmetallics with high modulus fibers with tension fatigue loads for 10^7 cycles [2.29, 2.48]

$$0.70 \leq \frac{FS}{TS} \leq 0.95. \tag{2.97}$$

However, for fully reversed stress [2.48, 2.50] for unidirectional and cross plied laminates

$$0.30 \le \frac{FS}{TS} \le 0.60. \tag{2.98}$$

Surface endurance strengths [2.32] such as in gear teeth with contact stresses are generally 2.5–5.0 times flexural endurance limits.

3. Low Cycle Fatigue Using Strain

The $\sigma_e^1 - N$ curve may also be presented as ε, strain, versus N, cycles. Since the yield is exceeded at 10^4 cycles the dimensionless value of strain, ε, is used. Fig. 2.32 a data curve by S. S. Manson et al. [2.67] shows various metals and condition on one ε versus N curve. Boller and Seegar [2.3] uses the same data but separate plots of each material are presented. Individual curves for each material may be constructed from handbook data *Fatigue Curves With Testing* by Peter Weihsmann, [2.59].

Note some authors [2.6, 2.35] present total strain amplitude versus $2N$ or a form of

$$\frac{\Delta \varepsilon_t}{2} = \frac{\Delta \varepsilon_e}{2} + \frac{\Delta \varepsilon_p}{2} = \frac{\sigma_f^1}{2}(2n)^b + \varepsilon_f^1(2n)^c \tag{2.99}$$

The curve Fig. 2.33 is constructed as follows

Point A $\Delta \varepsilon = (\sigma_{ut}/E)$ plotted at $N = 1/4(0.25)$
Point B $\Delta \varepsilon = (\sigma_{e'}/E)$ at stated cycles $10^6, 10^8, 5 \times 10^8$ but obtain N greater also include k_m Eqs. (2.89) and (2.90).
Point C $\Delta \varepsilon = \varepsilon_f = \ln e[100/(100 - R_a)]\, Ra - \%$ reduction of area plotted at $N = 1/4(0.25)$
Point D $\Delta \varepsilon = \varepsilon_e = (\sigma_{yt}/E)$ Plotted at $N = 10^4$ cycles the boundary between plastic and elastic strain

This method was used on the curve by Manson et al. [2.67] in Fig. 2.32 for A 356 aluminum casting and 17-4pH (H900) where both materials are different. The method showed good agreement in Fig. 2.34.

Manson [2.35] simplified the previous Eq. 2.99 to yield

$$\Delta \varepsilon = 3.5 \frac{\sigma_{ut}}{E} \frac{1}{N^{0.12}} + \varepsilon_f^{0.6} \frac{1}{N^{0.6}} = \Delta \varepsilon_e + \Delta \varepsilon_p \tag{2.100}$$

where

E – is Youngs Module
σ_{ult} – ultimate tensile strength
ε_f – true strain at fracture in tension

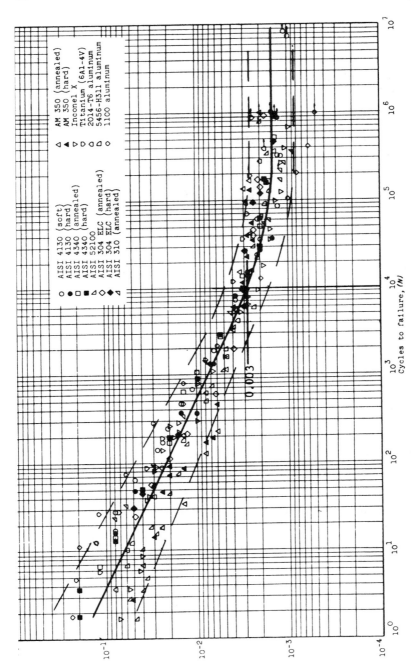

Figure 2.32 Effect of diametral strain range of fatigue [2.18.2.67]. Reprinted by permission John Wiley and Sons Inc.

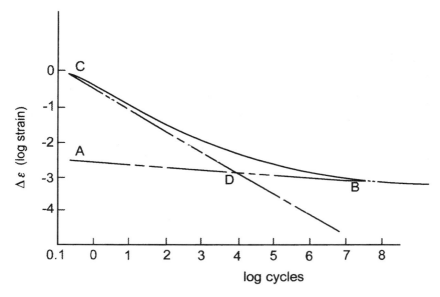

Figure 2.33 $\Delta\varepsilon$ versus 10 N cycles construction [2.59].

When calculating the variation of $\Delta\varepsilon$ versus N note the variables σ_{ult}, σ_{yt}, E, ε_f, estimates could be made on the $\Delta\varepsilon$ versus N curve for C_v coefficient of variation, from Fig. 2.34.

C. σ_r–σ_m Curves

The average design lines [2.65] and Fig. 2.11 are available but the exact spread about the lines is not given. The curves with the spread about the average design lines [2.27] require much labor to generate the curves alone for room temperature data. The ideal curves conditions are for

1. Fatigue curves like [2.27] are desirable but the heat treatment, sizes, surface finish and temperature variations are not mentioned.
2. The first curves [2.65] and Fig. 2.11 show no coefficient of variation or temperature variation. The coefficient of variation averages are known and the σ_e and $\breve{z}_{\sigma e}$ may be calculated and plotted.
3. Fatigue curves Fig. 2.35 have to be constructed from σ_e' and σ_{yt} and σ_{ult}. The variation with temperature may be developed if the mechanical properties as a function of temperature is known.

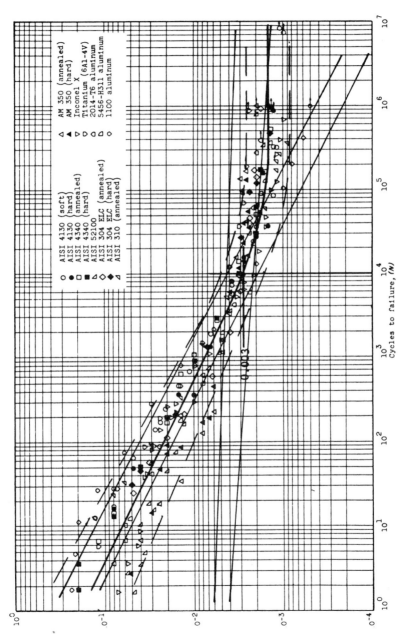

Figure 2.34 Effect of diametral strain range of fatigue [2.18,2.67] with a 356 aluminum casting and 17-4 PH(900). Reprinted by permission John Wiley and Sons Inc.

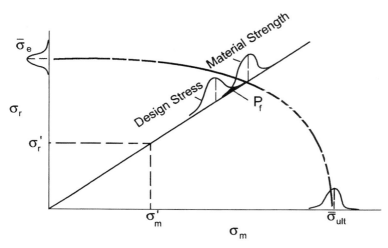

Figure 2.35 Fatigue curve construction.

4. Fatigue curves with the established method of using a factor of safety

  (3–4) – ductile materials

  (5–6) – brittle materials

with a $R(0.99)$ can also be drawn on a σ_r–σ_m curve Fig. 2.15 with many designs checked this way.

The conditions 1, 2, 3 are ideal for computer or handwritten solutions, when the σ_r'/σ_m' design line is known. Note in Fig. 2.35

$$\frac{\sigma_r'}{\sigma_m'} = \text{zero} \qquad \text{thus } \sigma_{yt} \text{ or } \sigma_{ult} \text{ is the intersection on the } \sigma_m \text{ line}$$

$$\frac{\sigma_r'}{\sigma_m'} = \text{infinite} \qquad \text{so } \bar{\sigma}_e \text{ is the intersection on the } \sigma_r \text{ line}$$

The equations for a one dimensional overlap calculation can be used by setting the p_f

$$P_f = \frac{1}{10^6} \text{ or } \frac{1}{10^3}$$

for a one sided distribution curve.

$$P_f = 9.8659 \times 10^{-10} \quad \text{for } z = -6 \text{ standard deviations}$$

$$\approx \frac{1}{10^9}$$

Solutions may be solved for only one unknown, area or a cross sectional dimension. The factor of safety can be calculated after solving for the unknown. Confidence levels can be found if the size of the input data is known for σ'_e, σ_{yt}, and σ_{ult}.

The one piece of information needed for the $\sigma_r - \sigma_m$ plot is the standard deviation $\check{z}_{\sigma e}$ of the endurance value σ_e from Eq. (2.55)

$$\sigma_e = k_a k_b k_c \cdots k_m \sigma'_e$$

This is a calculation similar to Example 2.16 which is a well behaved product

$$\check{z}_{\sigma e} = \sigma_e \left[C^z_{v\sigma'_c} + \sum C^2_{vk_i} \right]^{1/2} \tag{2.101}$$

EXAMPLE 2.18. Use the cantilever beam (Example 2.16) of a copper-based alloy

$\bar{\sigma}'_e = 0.35 s_{ult}$ use $s_{ult} = 80,000$ psi Fig. 2.28

$$C_{v\sigma'_e} = \frac{\sigma_{ult}/50}{0.35\sigma_{ult}} = \pm 0.0571 \tag{2.102}$$

$\check{z}_{\sigma'_e} = 0.0571 \bar{\sigma}'_e = C_{v\sigma'_e} \bar{\sigma}'_e$

k_m Eq.(2.89)10^{10} cycles

$\dfrac{\check{z}_m}{k_m} = \dfrac{0.0196}{0.85} = \pm 0.02305 = C_{vk_m}$

$\check{z}_m = 0.02305 \, k_m = C_{vk_m} k_m$

k_1 (Section A.12) radiation $k_l = 1$ $C_{vk_l} = 0 \therefore \check{z}_{kl} = 0$

k_k (Section A.11) dynamics (in stress calculation) $\bar{k}_k = 1$, $\check{z}_{k_k} = C_{vk_l} = 0$

k_j (Section A.10) fretting $\bar{k}_j = 0.75$, $\dfrac{\check{z}_j}{k_j} = \dfrac{0.0167}{0.75} = \pm 0.0222 = C_{vk_j}$

k_i (Section A.9) surface treatment none $k_1 = 1$ $C_{vk_j} = 0$

k_h Eq. (2.77) environment $k_h = 0.955$, $\dfrac{\check{z}_h}{k_h} = \dfrac{0.035}{0.955} = \pm 0.0367 = C_{vk_h}$

k_g (Section A.7) internal structure $k_g = 1$ $C_{vk_g} = 0$

k_f (Section A.8) residual stress, none $k_f = 1$ $C_{vk_f} = 0$

k_e (Section A.5) stress concentration $k_e = 1$ $C_{vk_e} = 0$

applied to stress

k_d (Section A.4) temperature (R.T.) $k_d = 1$ $\quad C_{vk_d} = 0$

k_c (Section A.3) Reliability $k_c = 1$ $\quad C_{vk_c} = 0$

k_b (Eq.(2.60)) size and shape $k_b = 0.85$ $\quad C_{vk_b} = 0.06$

k_a (Fig. 2.14) Table 2.4 surface condition $\dfrac{\breve{z}_a}{k_a} = \dfrac{0.0406}{0.819} = 0.0496 = C_{vk_a}$

Now $\bar{\sigma}_e = k_a k_b k_c \cdots k_m \bar{\sigma}'_e$

$$= (0.819)(0.85)(0.955)(0.75)(0.85)(0.35[80,000 \text{ psi}])$$

$$\bar{\sigma}_e = 11,867 \text{ psi} \tag{2.103}$$

$$\breve{z}_{\sigma_e} = \bar{\sigma}_e \left[C^2_{v\sigma'_e} + \sum_{i=a}^{i=m} C^2_{vk_i} \right]^{1/2}$$

$$= \bar{\sigma}'_e[(0.0571)^2 + (0.0496)^2 + (0.06)^2 + (0.0361)^2$$

$$+ (0.0222)^2 + (0.02305)^2]^{1/2}$$

$$\frac{\breve{z}_{\sigma e}}{\bar{\sigma}_e} = \pm 0.1081 \tag{2.104}$$

$$\breve{z}_{\sigma_e} = \pm 0.1081(11,867) = 1283 \text{ psi}$$

Note:

$$\frac{\breve{z}_{\sigma e'}}{\sigma'_e} = \pm 0.0571 \text{ for material alone} \tag{2.105}$$

Now to construct a $\sigma_r - \sigma_m$ curve knowing

$$\frac{\breve{z}_{\sigma e}}{\bar{\sigma}_e} = \pm 0.1081 \qquad \frac{\breve{z}_{ut}}{\bar{\sigma}_{ut}} = \pm 0.05 \qquad \text{Eq. (2.54)}$$

Construct the mean values on the line in Fig. 2.36

$$\frac{\sigma_m}{\sigma_{ut}} + \frac{\sigma_r}{\sigma_e} = 1$$

for a factor of safety 1.

The cantilever beam size is found from

$$\sigma_s = \sqrt{\sigma_r^2 + \sigma_m^2}$$

σ_s is plotted on the σ_r / σ_m curve solutions obtained per Sect. C.2

   5. Coupling Eq. (2.42)

$$t = -\frac{\sigma_S - \sigma_s}{\sqrt{\breve{z}_S^2 + \breve{z}_s^2}}$$

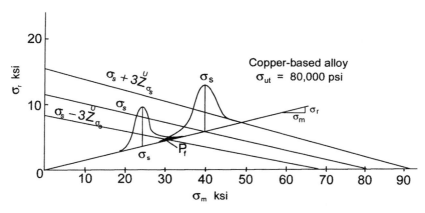

Figure 2.36 Coupling diagram for $\sigma_r - \sigma_m$ plot.

$$\text{set } P_f = \frac{1}{10^3}\frac{1}{10^6} \text{ or } (\frac{1}{10^9}, z = -6)$$

or some selected value from Table 2.3
then

$$\breve{z}_s = \left[\left(\frac{\partial\sigma_s}{\sigma'_r}\gamma_{\sigma'_r}\right)^2 + \left(\frac{\partial\sigma_s}{\partial\sigma'_m}\gamma_{\sigma'_m}\right)^2\right]^{1/2}$$

Remember σ_s and $\breve{z}_s$ have the unknown dimension, σ_m can be scaled from Fig. 2.36 as well as $\breve{z}_s$. This is now an algebra problem with C_v on each of the variables in the σ_s equation.

6. Monte Carlo
This is a computer solution where σ_S, $\breve{z}_{\sigma_S}$ are known and σ_s is known with a distribution on the variables be it Gaussian or Weibull. (σ_S can also be a Weibull distribution). The computer picks values on all known and unknown variables according to their distributions, finds all values of $\sigma_s > \sigma_S$ which is a failure, and does enough calculations to find

$$p_f = \frac{1}{\text{maximum calculations } 10^3, \ 10^6, \ 10^9}$$

should it find more than one failure in the first calculations the unknown dimension is made larger and the calculations repeated until a suitable answer is found.

7. Established method with a factor of safety N. Pick the $\sigma_s - 3\breve{z}_s$ line as design line and if $N = 4$ and take $1/4\,\sigma_S$ along the σ_r/σ_m line and this becomes a design point for σ_s. When the average value for the

unknown is found use the variations and calculate Eq. (2.42)

$$t = -\frac{\sigma_S - \sigma_s}{\sqrt{\breve{z}_S^2 + \breve{z}_s^2}}$$

$$N = \frac{\sigma_S}{\sigma_s}$$

then the actual P_f can be found for the design.

EXAMPLE 2.19. COMBINED FATIGUE STRESSES. A large 300-lb gear, Fig. 2.37 transmits torque in one direction through a shaft whose bearings are preloaded with a 100-lb force so that the load in the shaft varies from 0–200 lb. The critical area is the change in shaft diameter which is made of SAE 4340 steel. The two shaft diameters are required. Assume $D/d = 2$.

$$T_m = 45,000 \text{ in lb}$$
$$T_r = 15,000 \text{ in lb}$$
$$M_m = 150 \text{ lb } (30 \text{ in}) = 4500 \text{ in lb}$$
$$F_m = 100 \text{ lb}$$
$$F_r = 100 \text{ lb}$$

$$\tau_m = \frac{T_m r}{J} = \frac{16T_m}{\pi d^3}; \quad \tau_r = K_{tT}\frac{16T_r}{\pi d^3}; \quad \sigma_{xr} = K_{tB}\frac{M_m c}{I} = K_{tB}\frac{32M_m}{\pi d^3}$$

$$\sigma_{xm} = \frac{4F_m}{\pi d^2}; \quad \sigma_{xr} = K_{tF}\frac{4F}{\pi d^2}$$

From (2.9, 2.51, 2.54)

$$K_{tF} = 2.35 \quad K_{tB} = 2.025 \quad K_{tT} = 1.65 \quad C_v = \pm 0.1166 \text{ (Eq. 2.69)}$$

The stress concentration will be applied only to Eq. (2.49) σ_r since the system is considered ductile. The $\sigma_r - \sigma_m$ curve will be drawn and allowable stress

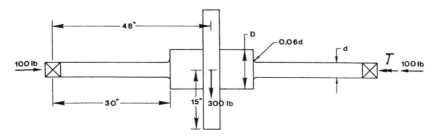

Figure 2.37 Bull gear and shafting.

levels obtained. The material parameters will be determined for $\check{z}_{\sigma_e}$ first. 4340 steel 1/2-in diameter $\sigma_{ut} = 210$ kpsi is heat treated and drawn to 800°F.

$$\sigma_e = k_a k_b k_c \ldots k_1 \sigma'_e \tag{2.106}$$

σ'_e Fig. 2.26.

$$\bar{\sigma}'_e = \frac{\sigma_{ut}}{2} \quad \check{z}_e = \frac{\sigma_{ut}}{30} \quad \frac{z'_e}{\bar{\sigma}'_e} = \frac{\sigma_{ut}}{30} \frac{2}{\sigma_{ut}} = \pm 0.0667 = C_{v\sigma'_e}$$

k_a Eq. (2.57) surface finish and Table 2.4

$$\bar{k}_a = 0.947 - 0.159 \times 10^{-5} \sigma_{ut} = 0.6131 \text{ with } \check{z}_a = \pm 0.0406 \text{ and}$$

$$C_v = \frac{\check{z}_a}{k_a} = \frac{0.0406}{0.6131} = 0.066$$

k_b (Eq. (2.60)) size shape $d < 2''$
$\quad \bar{k}_b = 0.85 \quad \check{z}_b = 0.06 \, \bar{k}_b \quad C_v = 0.06$
k_c Section A.3 Reliability coupling Eq. (2.42) $\qquad k_c = 1 \qquad C_v = 0$
k_d Section A.4 Temperature $k_d = 1$ since oil operates @ 180°F $\quad C_v = 0$
k_e Section A.5 Apply stress conditions on stresses $\quad k_e = 1 \qquad C_v = 0$
k_f Eq. 2.72 Shot peening $\check{z}_f = [(1.22 - 1.13)/3] = \pm 0.03, \; \bar{k}_f = 1.13,$
$\qquad\qquad\qquad\qquad\qquad\qquad\qquad\qquad\qquad\qquad C_v = 0.0265$
k_g Section A.7 Internal structure $\qquad\qquad\qquad\quad k_g = 1 \qquad C_v = 0$
k_h Section A.8 Environment (in oil) $\qquad\qquad\quad k_h = 1 \qquad C_v = 0$
k_i Section A.9 No surface treatment $\qquad\qquad\quad k_i = 1 \qquad C_v = 0$
k_j Section A.10 Fretting $\qquad\qquad\qquad\qquad\quad k_j = 0.95 \quad C_v = 0$
k_k Section A.11 Shock in stress calculation $\qquad k_k = 1 \qquad C_v = 0$
k_l Section A.12 Radiation $\qquad\qquad\qquad\qquad\quad k_l = 1 \qquad C_v = 0$
k_m Eq. 2.90 10^7 cycles $\qquad\qquad\qquad\qquad\quad k_m = 1 \qquad C_v = 0$

Evaluation per Example 2.18

$$\bar{\sigma}_e = (0.6131)(0.85)(1.13)(0.95)\left(\frac{210,000}{2} \text{ psi}\right)$$

$$\bar{\sigma}_e = 58,741 \text{ psi}$$

$$\check{z}_{\sigma_e} = \bar{\sigma}'_e \left[C^2_{v\sigma'_e} + \sum_{l=1}^{l=n} C^2_{vki} \right]^{1/2} \tag{2.107}$$

$$= \bar{\sigma}_e [(0.0667)^2 +) (0.066)^2 + (0.06)^2 + (0.0265)^2]^{1/2}$$

$$\frac{\check{z}_{\sigma_e}}{\bar{\sigma}'_e} = \pm 0.1145$$

$$\frac{\check{z}_{\sigma ut}}{\bar{\sigma}_{ut}} = \pm 0.05 \text{ (Eq. (2.54))}$$

The combined stresses for the shoulder are

$$\sigma'_r = \sqrt{\left[\sigma^2_{xr} + 3\tau^2_{xyr}\right]^{1/2}} \qquad (2.108)$$

where

$$\sigma_{xr} = K_{tB}\frac{32M_m}{\pi d^3} + K_{tF}\frac{4F_a}{\pi d^2} = 2.025\frac{32(4500 \text{ in lb})}{\pi d^3} + 2.35\frac{4(100 \text{ lb})}{\pi d^2}$$

$$\sigma_{xr} = \frac{16}{\pi d^3}[18,225 + 58.75d]$$

$$\tau_{cxyr} = K_{tT}\frac{16T_r}{\pi d^3} = \frac{16}{\pi d^3}[1.65(15,000 \text{ in lb})] = \frac{16}{\pi d^3}(24,750)$$

$$\sigma'_r = \frac{16}{\pi d^3}[[18,225 + 58.75d]^2 + 3[24,750]^2]^{1/2}$$

and

$$\sigma'_m = [\sigma^2_{xm} + 3\gamma^2_{xym}]^{1/2}$$

$$\sigma_{xm} = \frac{4F_m}{\pi d^2} = \frac{16}{\pi d^3}[25d] \qquad (2.109)$$

$$\tau_{cxym} = \frac{16T_m}{\pi d^3} = \frac{16}{\pi d^3}(45,000 \text{ in lb})$$

$$\sigma'_m = \frac{16}{\pi d^3}\sqrt{[(25d) + 3(45,000)^2]^{1/2}}$$

find the ratio σ'_r/σ'_m

$$\frac{\sigma'_r}{\sigma'_m} = \frac{\left[(18,225 + 58.75d)^2 + 3[24,750]^2\right]^{1/2}}{[(25d)^2 + 3(45,000)^2]^{1/2}} \qquad (2.110)$$

neglect the terms with d

$$\frac{\sigma'_r}{\sigma'_m} = 0.5976 = \frac{1}{1.67732}$$

Note the σ'_r/σ'_m line will be used with no variation assuming the angle variation is small.

Obtain a solution for Section C.7.

1. Mean Curve
    Using mean curve Fig. 2.38 and f.s. = 3 then $\sigma_s - 3\breve{z}_\sigma s$ curve
    a. mean curve f.s. = 3

$$\sigma'_r = 13,426 \text{ psi} = \frac{16}{\pi d^3}[(18,225)^2 + 3(24,750)^2]^{1/2} \quad (2.111)$$

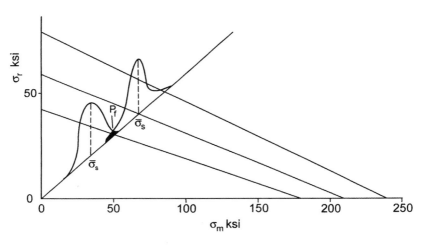

Figure 2.38 Bull gear material and stress due to loading.

solving for d^3

$$d^3 = \frac{46,582}{13,426} \frac{16}{\pi} = 17.670 \text{ in}^3$$

$$d = 2.605'' \, \phi \text{ [does not include sudden loading]}$$

b. Section C.4 $\sigma_s - 3\breve{z}_{\sigma_S}$

$\sigma_S - 3\breve{z}_{\sigma_S}$ curve with f.s. $= 3$ and with $\sigma'_r = 10,000$ psi

$$d^3 = \frac{46,582}{10,000} \frac{16}{\pi} = 23.724 \text{ in}^3 \tag{2.112}$$

$$d = 2.8734'' \phi \text{ [does not include sudden loading]}$$

c. Section C.5

Now obtain a solution for d using the coupling equation Eq. (2.42), or a Monte Carlo program could be used here. Here $z = -6$ standard deviations, $P_f = 9.8659 \times 10^{-10}$ from Section C.5 with the σ'_r Eq. (2.108) simplified by dropping d.

$$\sigma'_r = \left[\left(K_{tB} \frac{32M_m}{\pi d^3} \right)^2 + 3 \left(K_{tT} \frac{16T_r}{\pi d^3} \right)^2 \right]^{1/2}$$

and

$$\sigma'_m = \frac{16T_m}{\pi d^3} \sqrt{3}$$

now

$$\sigma_s = [(\sigma_m')^2 + (\sigma_r')^2]^{1/2}$$

$$= \frac{16}{\pi d^3}\left[\left(\sqrt{3}T_m\right)^2 + \{(K_{tB}2M_m)^2 + 3(K_{tT}T_r)^2\}\right]^{1/2} \qquad (2.113)$$

evaluating for average values

$$\bar{\sigma}_s = \frac{16}{\pi d^3}[(\sqrt{3}[45{,}000 \text{ in lb}]^2 + \{(2.025[2 \times 4500 \text{ in lb}])^2$$
$$+ 3(1.65[15{,}000 \text{ in lb})^2\}]^{1/2}$$

$$\bar{\sigma}_s = \frac{16}{\pi d^3}(90{,}802)$$

2. Card Sort

We have six variables in Eq. (2.113) to generate a maximum value for a card sort and find $\breve{z}_s$ from Example 2.14, for $3\,\breve{z}$ measuring errors 0.01, Eq. (2.69), and with

$$\breve{z}_d = 0.010\,\bar{d}.$$

yielding

$$d_{\text{mind}} = [1 - 3(0.01)]\bar{d} \quad (T_m)_{\max} = 1.01T_m \quad (K_{tB})_{\max} = 1.3498\bar{K}_{tB}$$
$$= 0.970\bar{d} \qquad\qquad (T_r)_{\max} = 1.01T_r$$
$$\qquad\qquad (M_r)_{\max} = 1.01M_r \quad (K_{tT})_{\max} = 1.3498\bar{K}_{tT}$$

Then from Fig. 2.4, and Table 2.2 with

$$X\,\breve{z}_s = 7.593\,\breve{z}_s = \sigma_{s\,\max} - \bar{\sigma}_s \qquad (2.114)$$

Substituting into Eq. (2.113)

$$\sigma_{s\,\max} = \frac{16}{\pi d^3}107{,}425$$

Evaluating

$$\breve{z}_s = \frac{\dfrac{16}{\pi d^3}(107{,}425 - 90{,}801)}{7.593} = \frac{16}{\pi d^3}2{,}189.40$$

$$\frac{\breve{z}_s}{\bar{\sigma}_s} = \frac{\dfrac{16}{\pi d^3}2189.40}{\dfrac{16}{\pi d^3}90{,}801} = 0.02411(2.411\%)$$

So

$$\breve{z}_s = \pm 0.02411\bar{\sigma}_s$$

Now to substitute into the coupling equation

$$t = -\frac{\sigma_S - \sigma_s}{\sqrt{\breve{z}_s^2 + \breve{z}_s^2}} = -6 \text{ [six sigma criterion]} \tag{2.115}$$

Scale from Fig. 2.38 $\sigma_S = \dfrac{2.18}{1.8} \, 50{,}000 = 78{,}056$ psi. Then from the spread

$$\breve{z}_S = \frac{1.45}{1.80} \, 50{,}000 \frac{1}{6std} = 6{,}713 \text{ psi}$$

substituting

$$6 = \frac{78{,}056 - \bar{\sigma}_s}{[(6{,}713)^2 + (0.02411\bar{\sigma}_s)^2]^{1/2}}$$

so both sides

$$36 = \frac{(78{,}056)^2 - 2(78{,}056)\sigma_s + \sigma_s^2}{(6{,}713)^2 + (0.02411\sigma_s)^2}$$

Combining

$$0.979073 \, \sigma_s^2 - 2(78{,}056) \, \sigma_s + [(78{,}056)^2 - 36(6713)^2] = 0$$

$$A \, \sigma_s^2 + B\sigma_s + C = 0$$

$$\sigma_s = \frac{-B \pm \sqrt{B^2 - 4Ac}}{2A}$$

[safety device p_f high]

$$\sigma_{s2} = \frac{2(78{,}056) + 82{,}846}{2(0.979073)} = 122{,}033 \text{ psi}$$

[Structural member no dynamic effects from loading]

$$\bar{\sigma}_{s1} = \frac{2(78{,}056) - 82{,}846}{2(0.97073)} = 37{,}416$$

Now

$$\bar{\sigma}_{s1} = 37{,}416 = \frac{16}{\pi d^3}(90801)$$

solve for d^3

$$d^3 = \frac{16}{\pi}\left(\frac{90{,}801}{37{,}016}\right) = 12.3596$$

$$d = 2.312$$

With no sudden loading taken into consideration

Safety factor

$$N = \frac{\sigma_S}{\sigma_s} = \frac{78,056}{37,416} = 2.0862 \quad 3 = N \quad \text{safety factor on other solutions}$$

The gear will be mounted on the shaft and

1. It may be keyed on the shaft with one or more keys
2. A spline may be the more efficient method with locknuts etc.
3. The gear could be flanged and the shaft mounted to it. This interface should be checked.
4. The shock load should be incorporated. Complete the back checking on sizes
5. Calculate a critical speed and include any base vibration motions
6. Include the axial force and check the factor of safety.

D. Fatigue Considerations in Design Codes

National codes and standards frequently provide methods of analysis that address the problem of fatigue, or cyclic loading. In applications where use of such codes is mandatory, the rules must be followed in detail. A notable example of this is provided by the ASME Boiler and Pressure Vessel Code [2.61] which contains detailed procedures for evaluating the fatigue behavior of pressure vessels and pressure vessel parts. This is an excellent treatment of the subject and is recommended to the reader for study and use (see also [2.56]). Information on the American Institute of Steel Construction Code (AISC) is presented in [2.51,2.64]. An excellent source of fatigue data for aircraft materials is also presented in [2.65].

In the example in Fig. 2.37 allowable stress can be developed for comparison. If the impact constant k_k is placed in the stress calculations, the allowable stress is 0.31 σ_{ut}.

E. Summary For Fatigue Calculations

1. When designing a system or a component, the critical frequencies, deflections, shock and vibration levels, operating temperatures, and surrounding environment must be known. The system must meet the overall requirements while often components must exceed them, as in deflections. Components with maximum deflections A when assembled will always have larger maximum deflections, as do springs in series.

2. Determine the component critical parameters such as deflections, stresses, frequency, or failure modes. Attempt to assign a value to these parameters even though it is understood that they will often change.

3. When the components are designed, the maximum and minimum static loads should be corrected to include the shock and vibration effects in the design. This does not mean that a vibration analysis can be neglected. Also, fracture mechanics should be checked to include important parameters.

4. Free-body diagrams when drawn will determine end conditions at mating parts. Self-aligning bearings develop simply supported ends while double bearings approach fixed-end conditions.

5. The determination of the effects of k values, Eq. (2.55), on endurance limits will force the designer to select the materials and/or the manufacturing process such as machined or cast parts, heat treatment, and surface coating or treatment.

6. When the $\sigma_r - \sigma_m$ curve is selected or drawn, the proper safety factor should be determined.

7. When complex systems or parts are designed, a finite element check for frequencies, deflections, and loads from the known operating conditions will allow a check on a system design before parts are fabricated. The parts should further be checked using fracture mechanics for acceptability.

8. When a part or a system is manufactured, finished products should be tested to failure to verify the designing and manufacturing. However, there are cases where only one item is produced and used. Then, some form of non-destructive inspections at servicing must be performed to maintain a check on the system.

9. Keep a record of successful designs and failures so that design criteria can be modified for a particular class of designs. This is how design criteria and future design codes will be developed.

EXAMPLE PROBLEM 2.20. An Egiloy metal pulley belt Fig. 2.39 is driven by a sprocket metal pulley of radius r. The ultimate strength is 368,00 psi; $\sigma_{yt} = 280,000$ psi; and $\sigma'_e = 128,800$ psi. The starting torque of 5 oz-in plus bending of the belt around the pulley creates the stress

$$\sigma_t = F/A = \frac{5}{r}\frac{1}{16\frac{\text{oz}}{\text{lb}}}\frac{1}{(0.5333 \text{ in})(0.0015 \text{ in})} \tag{2.116}$$

$$\sigma_t = \left(\frac{391}{r}\right) \text{ psi}$$

The bending stress due to bending around the pulley is

$$\frac{1}{r} = \frac{d^2y}{dx^2} = \frac{M}{EI}$$

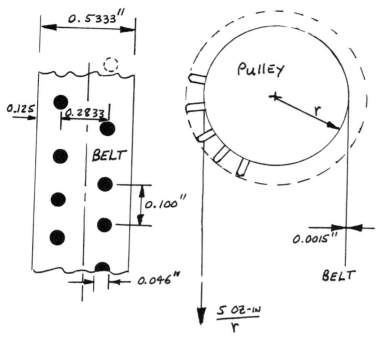

Figure 2.39 Example 2.20 metal belt on a pulley.

and

$$\sigma_B = \frac{Mc}{I} = \frac{E}{r}c \tag{2.117}$$

with $c = 0.0015\,\text{in}/2$ and $E = 30 \times 10^6$ psi

$$\sigma_B = \frac{30 \times 10^6}{r}\left(\frac{0.0015}{2}\right) = \left(\frac{22{,}500}{r}\right)\text{ psi}$$

$$\sigma_{\text{mean}} = \frac{\sigma_{\text{max}} + \sigma_{\text{min}}}{2} = \frac{1}{2}\left[\left(\frac{22{,}500}{r} + \frac{391}{r}\right) + \frac{391}{r}\right]$$
$$= \left(\frac{11{,}641\text{ psi}}{r}\right) \approx \frac{Ec}{2r} \tag{2.118}$$

$$\sigma_r = \frac{\sigma_{\text{max}} - \sigma_{\text{min}}}{2} = \frac{1}{2}\left[\left(\frac{22{,}500}{r} + \frac{391}{r}\right) - \frac{391}{r}\right]$$
$$= \left(\frac{11{,}250}{r}\text{ psi}\right) \approx \frac{Ec}{2r} \tag{2.119}$$

Find k_t for the belt holes [2.9] using the tension curve where $h \leq 20d$ or $h \leq 20(0.046) \leq 0.920''$

$$\frac{d}{b} = \frac{0.046}{0.125} = 0.368 \quad K_t \approx 3.21 \quad t/r = \frac{0.0015}{0.046''/2} = 0.065 \ll 3$$

From Fig. 2.16 and Table 2.7 $q \approx 0.96$ QT steel. Now find $\check{z}_f$ with $K_f = 1 + q(K_t - 1)$ The

$$C_q = \frac{\check{z}_q}{q} = \pm 0.0833 \text{ Table 2.7 } q \approx 0.96$$

$$C_{K_t} = \frac{\check{z}_t}{\bar{K}_t} = \pm 0.109 \ K_t \approx 3.21 \text{ Sect. A.5}$$

$$\frac{\check{z}_{K_f}}{K_f} = \frac{\left[\left(\frac{\partial K_f}{\partial q} \check{z}_q \right)^2 + \left(\frac{\partial K_f}{\partial K_t} \check{z}_{kt} \right)^2 \right]^{1/2}}{[1 + q(K_t - 1)]} \times 100 \tag{2.120}$$

with parameters from Section A.5 Example 2.17

$$\frac{\partial K_f}{\partial q} = (\bar{K}_t - 1) \text{ and } \frac{\partial K_f}{\partial K_t} = \bar{q}$$

$$\check{z}_q = 0.0833\bar{q} \qquad \check{z}_t = 0.109 K_t$$

We find

$$\frac{\check{z}_f}{K_f} = \pm 12.16\%$$

Now to develop the $\sigma_r - \sigma_m$ plot by establishing Eq. (2.55) and S_{ut}

$$\sigma_e = k_a k_b k_c \cdots \sigma_e' \tag{2.121}$$

k_a–ground finish Eq. (2.56) $\check{z}_a = \pm 0.103$ Table 2.4

$$k_a = 0.7429$$

$$C_{v_a} = \frac{\check{z}_a}{k_a} = \frac{0.103}{0.7429} = 0.1386$$

k_b–size effect in $\sigma_e' = 128{,}800 \ k_b = 1$ $\check{z}_b = C_{vb} = 0$
k_c – is 1 when solving for p_f $\check{z}_c = C_{v_c} = 0$
k_d – 1 since temperature near room temperature $\check{z}_d = C_{v_d} = 0$
k_e – use K_f with stress $k_e = 1$ here and $\check{z}_e = C_{v_e} = 0$
k_f – $0.0015''$ thk tape $k_f = 1$ $\check{z}_f = C_{v_f} = 0$

k_g – 1 for what is known $\qquad \check{z}_g = C_{v_g} = 0$
k_h – corrosion neglect $k_h = 1$ $\qquad \check{z}_h = C_{v_h} = 0$
k_i – surface treatment none $k_i = 1$ $\qquad \check{z}_i = C_{v_i} = 0$
k_j – fretting none $k_j = 1$ $\qquad \check{z}_j = C_{v_j} = 0$
k_k – shock or vibration, this is a smooth
  operation $k_k = 1$ $\qquad \check{z}_k = C_{v_k} = 0$
k_l – radiation none is known $k_l = 1$ $\qquad \check{z}_l = C_{v_l} = 0$
σ'_e – 128,000 $\qquad \check{z}_{\sigma'_e} = \pm 0.08\sigma'_e$

In addition $k_m = k_{life}$ from (2.2) for up to 10^{10} cycles, $K_m = 0.85$ Eq (2.89)

$$C_{vm} = \frac{0.01967}{0.85} = \frac{\check{z}_m}{k_m} = 0.0231$$

now find Eq. (2.55)

$$\bar{\sigma}'_e = (0.7429)128,000(0.85)$$
$$\bar{\sigma}'_e = 80,828 \text{ psi}$$

$$\check{z}_{\sigma'_e} = \sigma'_e \left[C_v^2 + \sum_{i=1}^{i=n} C_{vhi}^2 \right]^{1/2} \tag{2.122}$$

$$\frac{\check{z}_{\sigma'_e}}{\sigma'_e} = \left[(0.08)^2 + (0.1386)^2 + (0.0231)^2 \right]^{1/2} = \pm 0.161$$

with Table 2.1

$$\frac{\check{z}_{ult}}{S_{ult}} = \pm 0.05$$

A $\sigma_r - \sigma_m$ plot can be drawn

$$\check{z}_{\sigma'_e} = \pm 0.161(80,828 \text{ psi}) \text{ and } \check{z}_{ult} = \pm 0.05(368,000)$$

The

$$\frac{\sigma_r}{\sigma_m} = \frac{K_f \dfrac{11,250}{r}}{K_f \dfrac{11,641}{r}} = 0.9664 \approx (1)$$

K_f used both in σ_r, σ_m as $s_{ult} > 200,000$ psi

The S and $\check{z}_S$ values are derived from Fig. 2.40. The s value

$$s^2 = \sigma_m^2 + \sigma_a^2 = \left(\frac{1}{2} \frac{Ec}{r} \right)^2 + \left(\frac{1}{2} \frac{Ec}{r} \right)^2$$

$$s = \frac{\sqrt{2}}{2} \frac{Ec}{r} \tag{2.123}$$

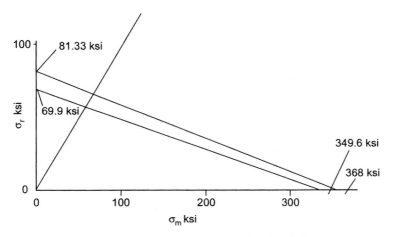

Figure 2.40 $\sigma_r - \sigma_m$ curve for Egiloy metal pulley belt.

$$\breve{z}_s = \frac{\sqrt{2}}{2}\frac{Ec}{2}\left[\breve{z}_c^2 + \breve{z}_E^2 + \breve{z}_r^2\right]^{1/2}$$

$$c = \frac{0.0015}{2} \qquad E = 30 \times 10^6 \qquad r = \text{unknown Example 2.14}$$

$$\frac{\breve{z}_c}{c} = \frac{\pm 0.0001/3}{0.0015} = 0.0222 \qquad \frac{\breve{z}_E}{E} = \pm 0.03 \qquad \frac{\breve{z}_r}{r} = \pm 0.5\% = 0.005\bar{r}$$

$$\breve{z}_c = 0.0222c \qquad \breve{z}_E = \pm 0.03E \qquad \breve{z}_r = \pm 0.005\bar{r}$$

$$\breve{z}_s = \frac{\sqrt{2}}{2}\frac{Ec}{r}[(0.03)^2 + (0.0222)^3 + (0.005)^2]^{1/2} \tag{2.124}$$

$$\frac{\breve{z}_s}{s} = \pm 0.0377 \tag{2.125}$$

Using the coupling equation Eq. (2.42) along the slanted line.

$$t = \frac{S-s}{\left[\sqrt{\breve{z}_S^2 + \breve{z}_s^2}\right]} = \frac{96,000 - s}{[(15,000 \text{ psi})^2 + (0.0377s)^2]^{1/2}} \tag{2.126}$$

$$t^2 = \frac{(96,000 - s)^2}{(15,000)^2 + 1,4213 \times 10^{-3}s^2}$$

$$(96,000)^2 - 2(96,000)s + s^2 = (15,000t)^2 + 1.4213 \times 10^{-3}(ts)^2$$

$$[1 - 1.42131 \times 10^{-3}t^2]s^2 - 2(96,000)s + [(96,000)^2 - (15,000t)^2] = 0$$

$$s = \frac{-B \pm [B^2 - 4AC]^{1/2}}{2A}$$

Condition 1. Table 2.3

$$P_f = \frac{1}{10^6} \qquad t = -4.7534$$

$$s = \frac{96,000 \pm 72,226}{0.96789}$$

$s_2 = 173,806$ psi safety device $s_1 = 24,563$ structural member

Condition 2. Table 2.3

$$P_f = \frac{1}{100} \qquad t = -2.3263$$

$$s = \frac{96,000 \pm 35,765}{0.99231}$$

$s_2 = 132,786$ psi safety device $s_1 = 60,702$ structural member

Now

$$s = s_1 = \frac{\sqrt{2}}{2} = \frac{Ec}{r}$$

$$P_f = \frac{1}{100}$$

$$r = \frac{\sqrt{2}}{2}\frac{Ec}{s_1} = \frac{\sqrt{2}}{2}\frac{30 \times 10^6 \dfrac{0.0015}{2}}{60,702 \text{ psi}} = 0.26210$$

$$2r = d = 0.5242'' \ \phi$$

$$P_f = \frac{1}{10^6}$$

$$r = \frac{\sqrt{2}}{2}\frac{30 \times 10^6 \dfrac{0.0015}{2}}{24,563} = 0.64772$$

$$2r = d = 1.2954''$$

This curve is shown in Fig. 2.41.

F. Monte Carlo Fatigue Calculations

The computer fatigue calculations using the Gaussian and Weibull distributions are set up as in Fig. 2.6 with t (Eq. (2.42)) set to obtain low probability of failure. The computer randomly picks a value for the unknown variable and calculates the load or stress while for the capacity or strength of a part or material the computer picks a value with in the distribution found previously. If the capacity is greater than the load or stress the selection is good, however, if load or stress is greater than

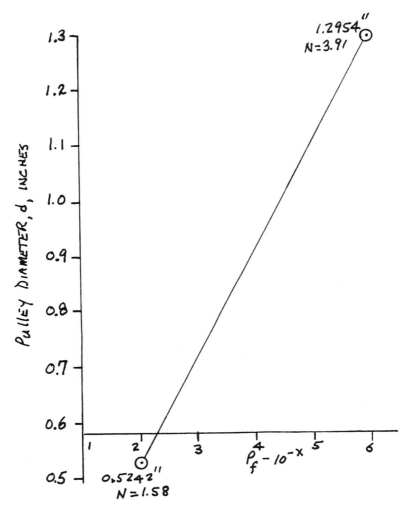

Figure 2.41 P_f, structural member, versus pulley diameter for 0.0015″ thick egiloy belt for 10^{10} cycles.

capacity, this is a failure. The selected value of t can set a P_f of 10^{-6} or one failure in a million calculations of material capacity selections. A properly programmed computer converges to the correct value of the unknown variable. Appendix B presents the set up for the two examples 2.21 and 2.22.

EXAMPLE 2.21. Use a tip loaded cantilever beam, Example 2.16, with a radius at the wall of r/h equal to 0.01 and with w/h, beam width to depth equal to 6 which yields a k_t of 2.10. Also from Example 2.16 using

$$\frac{\overset{\smallsmile}{z}_\sigma}{\bar{\sigma}} = \pm 0.02674 \tag{2.127}$$

with

$$\bar{\sigma} = \frac{6\bar{P}\bar{L}}{\bar{b}\bar{h}^2}$$

The Monte Carlo simulation is used to solve for a $P_f = 10^{-6}$ for the beam cross section. The beam stress $\bar{\sigma}$ is expressed from Gaussian parameters. The aluminum casting is expressed as a Gaussian and Weibull distribution from Example 1.4 with

Weibull Averages
β (shape) – 1.561
δ (scale) – 3766.9
γ (threshold) – 43,109 psi

Gaussian Averages
μ (mean) – 46,508 psi
$\overset{\smallsmile}{z}$ (standard deviation) – 2159 psi

The first solution is a Weibull distribution for strength and a Gaussian distribution for the stress.

Weibull formulation. The curve $\sigma_r - \sigma_m$ (Fig. 2.42) is set up like Figs 2.38 and 2.40. The end of the failure line on the σ_m axis is the lowest value

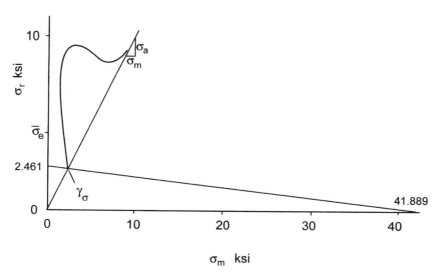

Figure 2.42 γ_σ Weibull aluminum casting parameter.

of γ from the Weibull (Example 1.4) while the lowest value of γ on the σ_r axis is taken from the low side of the known Gaussian representations. This is developed later (Eq. (2.131)). The σ_e values on the σ_r axis are developed from Eq. (2.55) and Example 2.18.

$$\check{z}_{\sigma e} = \bar{\sigma}'_e \left[C^2_{v\sigma e'} + \sum_{i=1}^{i=1} C^2_{v_{ki}} \right]^{1/2} \tag{2.128}$$

The value for $\bar{\sigma}_e$ for $\bar{\gamma} = 43,109$ psi Fig. 2.29 is 7500 psi and $\check{z}_{\sigma'_e}$ is roughly $2500/3$.

The coefficient of variation is

$$C_{v'_e} = \frac{\check{z}_{\sigma'_e}}{\bar{\sigma}'_e} = 0.111$$

The rest of the variations for the ki's are as follows.

k_m Eq (2.90) for 5×10^8 cycles	$\bar{k}_m = 1$	$C_{vm} = 0$
k_1 Section A.12 radiation	$\bar{k}_1 = 1$	$C_{vl} = 0$
k_h Section A.12 dynamics with stresses	$\bar{k}_h = 1$	$C_{vh} = 0$
k_j Section A.10 fretting (none)	$\bar{k}_j = 1$	$C_{vj} = 0$
k_i Section A.9 surface treatment (none)	$\bar{k}_i = 1$	$C_{vi} = 0$
k_h Eq. (2.80) environment	$\bar{k}_h = 0.88$ $C_{vh} = \frac{0.040}{0.88} = 0.0455$	
k_g Section A.7 internal structure	$\bar{k}_g = 1$	$C_{vg} = 0$
k_f Section A.6 residual stress (none)	$\bar{k}_f = 1$	$C_{vf} = 0$
k_e Section A.5 stress concentration applied to stress Eqs.	$\bar{k}_e = 1$	$C_{ve} = 0$
k_d Section A.4 room temperature	$\bar{k}_d = 1$	$C_{vd} = 0$
k_c Secton A.3 reliability	$\bar{k}_c = 1$	$C_{vc} = 0$
k_b Eq. 2.60 $b \leq 2''$ size and shape	$\bar{k}_b = 0.85$	$C_{vb} = 0.06$
k_a Fig. 2.14 and Table 2.4 mill scale hot rolled	$\bar{k}_a = 0.80$	$C_{va} = 0.11$

Now substituting $\bar{k}'_i$s Eq. (2.55)

$$\bar{\sigma}_e = \bar{k}_a \bar{k}_b \cdots \bar{k}_m \bar{\sigma}'_e$$
$$\bar{\sigma}_e = (0.8)(0.85)(0.88)7500 \text{ psi} \tag{2.129}$$
$$\bar{\sigma}_e = 4,488 \text{ psi}$$

Now substituting into Eq. (2.128)

$$\check{z}_{\sigma e} = \bar{\sigma}'_e[(0.111)^2 + (0.11)^2 + (0.06)^2 + (0.0455)^2]^{1/2}$$
$$\check{z}_{\sigma e} = \pm 0.1735 \bar{\sigma}_e = C_{v\bar{\sigma} e} \bar{\sigma}'_e \tag{2.130}$$

Note the variation on the $\bar{\sigma}'_e$ value is 0.111 compare to 0.1735 with all variations included.

The next step is to pick a γ for the σ_r axis from the Gaussian represented σ_e. The following is used

$$(\sigma_e)_L = \gamma = \sigma_e - 2.576\breve{z}_{\sigma e}$$
$$= \bar{\sigma}_e[1 - 2.576(0.1735)] \tag{2.131}$$
$$(\sigma_e)_L = \gamma = 2{,}482 \text{ psi}$$

Now the material failure line ends are defined. Next to define the stress, "s" along the σ_r, σ_m slope of 1, then set the material value for S, similar to Fig. 2.38 for interations using a Monte Carlo simulation can be performed.

In Example 2.16 stress in the beam was found

$$\sigma = \frac{6PL}{bh^2}$$

When the load P is varied from 0 to 30 lbs and back there are two components formed on the $\sigma_r - \sigma_m$ curve

$$\sigma_a = \frac{\sigma_{\max} - \sigma_{\min}}{2} = \left[\frac{6P_{\max}L}{bh^2} - \frac{6(0)L}{bh^2} \right] \frac{1}{2}$$
$$\sigma_a = \frac{1}{2}\sigma_{\max}$$

The same can be said for σ_m

$$\sigma_m = \frac{\sigma_{\max} + \sigma_{\min}}{2} = \frac{1}{2}\sigma_{\max}$$

Now the stresses have K_t multiplied by each of them. This is because a casting is sensitive to stress concentration in both σ_a and σ_m. Now s is along the line σ_a/σ_m equal to one. As in Fig. 2.38 s is required.

$$s = [\sigma_a^2 + \sigma_m^2]^{1/2}$$
$$s = K_t \frac{6P_{\max}L}{2bh^2} \sqrt{2} \tag{2.132}$$

Now $h = 2b$ so

$$s = \frac{6K_t P_{\max}L}{2b(2b)^2} \sqrt{2}$$
$$s = 1.06066 \frac{\bar{K}_t \bar{P} \bar{L}}{\bar{b}^3} \tag{2.133}$$

where

$$\frac{\overset{\vee}{z}_{kt}}{\bar{K}_t} = \pm 0.01163 \quad \frac{\overset{\vee}{z}_{\bar{P}}}{\bar{P}} = \pm 0.01 \quad \frac{\overset{\vee}{z}_L}{\bar{L}} = \pm 0.01 \quad \frac{\overset{\vee}{z}_b}{\bar{b}} = \pm 0.01$$

$$\bar{K}_t = 2.10 \quad \bar{P} = 30 \text{ lbs} \quad \bar{L} = 15'' \quad \bar{b} = \text{solve for}$$

This is the Gaussian stress formulation to obtain S from the Weibull failure line

$$\frac{1}{N} = \frac{\sigma_m}{\gamma_L} + \frac{\sigma_a}{(\gamma_{\sigma e})_L} \tag{2.134}$$

where

N – one for calculations

$\dfrac{\sigma_a}{\sigma_m} = 1$

γ_L – is 41,889 psi low side Eq. (1.74)

$(\gamma\sigma_e)_L$ – is 2,482 psi Eq. (2.131)

substituting

$$\frac{1}{1} = \sigma_m \left[\frac{1}{41,889} + \frac{1}{2,482} \right]$$

$$\sigma_m = 2343 \text{ psi} = \sigma_a$$

$$\gamma_\sigma = [\sigma_a^2 + \sigma_m^2]^{1/2}$$

$$\gamma_\sigma = \sqrt{2}\sigma_m = 3314 \text{ psi}$$

The parameters for the Weibull to obtain a conservative b.

$$\gamma_\sigma = 3314 \text{ low value for line intercept } \sigma_a/\sigma_m = 1 \tag{2.135}$$

$$\beta = 1.1299 \text{ Eq. (1.71) skews material property lower} \tag{2.136}$$

$$\theta = 4878.47 \text{ Eq. (1.72) larger to make } b \text{ larger} \tag{2.137}$$

Equations (2.133) and (2.135), (2.137) are used in a Monte Carlo simulation to obtain a solution for $\bar{b}$. This is discussed in Appendix B.

The second solution is that of Gaussian stress versus Gaussian fatigue strength.

Gaussian. The $\sigma_r - \sigma_m$ curve Fig. 2.43 is set up similar to Figs. 2.38 and 2.40 where the ultimate tensile strength μ and $\overset{\vee}{z}$ are used on the σ_m axis to obtain a

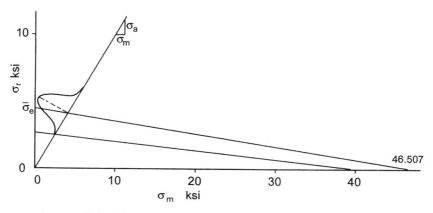

Figure 2.43 $\bar{S}, \check{Z}_S$ Gaussian aluminum casting parameters.

minimum

Eq. (1.68) $\mu_{\min} = 46{,}329$ psi

Eq. (1.69) $\check{z} = 3{,}038$ psi

so S_L

$$S_L = \mu_{\min} - 2.576\,\check{z}_{\max}$$

$$S_L = 38{,}503 \text{ psi with } \bar{S} = 46{,}507 \text{ psi} \tag{2.138}$$

Then $\bar{\sigma}_e$ is 4,488 psi from Eq. (2.129)

The mean line using $\bar{S} = 46{,}507$ and $\bar{\sigma}_e = 4488$ psi is used on the failure line with $\sigma_a/\sigma_m = 1$ and $N = 1$ in Fig. 2.43.

$$\frac{1}{N} = \left[\frac{\sigma_m}{46{,}507} + \frac{\sigma_a}{4488}\right]$$

$$\sigma_m = 4093 \text{ psi} = \sigma_a \tag{2.139}$$

$$\bar{\sigma}_S = [\sigma_a^2 + \sigma_m^2]^{1/2} = 5788 \text{ psi}$$

for the lower line at under $2.576\,\check{z}$ Fig. 2.43 $\dfrac{\sigma_{aL}}{\sigma_{mL}} = 1$ and $N = 1$

$$\frac{1}{N} = \left[\frac{\sigma_{mL}}{S_L} + \frac{\sigma_{aL}}{(\sigma_e)_L}\right]$$

$$1 = \frac{\sigma_{mL}}{38{,}503} + \frac{\sigma_{aL}}{2{,}482} \tag{2.140}$$

$$\sigma_{mL} = \sigma_{aL} = 2{,}332$$

$$\sigma_{SL} = [\sigma_{mL}^2 + \sigma_{aL}^2]^{1/2} = 3298 \text{ psi}$$

Now to obtain $\check{z}_S$

$$2.576\,\check{z}_S = \bar{\sigma}_S - \sigma_{SL}$$
$$= 5788 - 3298 \text{ psi} \tag{2.141}$$

$$\check{z}_S = 967 \text{ psi standard deviation} \tag{2.142}$$

and

$$\bar{S} = 5788 \text{ psi Gaussian mean} \tag{2.143}$$

The stress due to the load is Eq. (2.133).

Again a Monte Carlo simulation yields a value for $\bar{b}$ using Eqs (2.133), (2.142) and (2.143). The P_f is 10^{-6}. The Monte Carlo simulation is discussed in Appendix B. The computer results for iterations starting from the low side, are
Weibull

$$\bar{b} = 0.7015 \text{ in} \tag{2.144}$$

$$b_{max} = \bar{b}[1 + 2.576\,(0.01)] \tag{2.145}$$

$$b_{max} = 1.02576\,\bar{b}$$

$$b_{max} = 0.7196$$

$$\bar{h} = 2\bar{b} = 1.4030 \tag{2.146}$$

$$h_{max} = 2b_{max} = 1.43914 \text{ in} \tag{2.147}$$

The factor of safety is

$$N = \frac{S(50\%)}{\bar{s}} \tag{2.148}$$

$S(50\%)$ is derived from Eq. (1.18) where x is $S(50\%)$ when

$$p_f = 0.5 = \exp\left[-\left(\frac{x-\gamma}{\theta}\right)^{\beta}\right]$$

The natural log of both sides is

$$-0.693147 = -\left(\frac{x-\gamma}{\theta}\right)^{\beta}$$

The $1/\beta$ root of both sides yields with $\beta = 1.1299$

$$\left(\frac{x-\gamma}{\theta}\right) = 0.722978$$

Then
$$S(50\%) = x = 0.7229978\theta + \gamma \tag{2.149}$$

The parameters used
$$\gamma = 3314 \text{ psi} \quad \theta = 4878.47 \quad \beta = 1.1299$$
$$S(50\%) = 6{,}841 \text{ psi}$$

Now $\bar{s}$ is Eq. 2.133 with $\bar{b} = 0.7015$ in Eq. (2.144)
$$\bar{s} = 1.06066 \frac{\bar{K_t}\bar{P}\bar{L}}{b^3} \tag{2.150}$$
$$\bar{s} = 2{,}904$$

Substituting into in Eq. (2.148) with Eq. (2.149), (2.150).
$$N = \frac{6{,}841 \text{ psi}}{2{,}904} \tag{2.151}$$
$$N = 2.36$$

These Weibull cross section dimensions and safety factors are compared in Table 2.12 with the Gaussian solutions. The Gaussian computer result is for iterations starting from the low side.

Gaussian
$$\bar{b} = 0.875 \text{ in} \tag{2.152}$$

$$b_{\max} = 1.02576 \, \bar{b}$$
$$b_{\max} = 0.8975 \tag{2.153}$$

$$\bar{h} = 2\bar{b} = 1.750'' \tag{2.154}$$

$$h_{\max} = 2b_{\max} = 1.795 \text{ in} \tag{2.155}$$

The factor of safety N is
$$N = \frac{\bar{S}}{\bar{s}} \tag{2.156}$$

Table 2.12 Cross sectional dimensions Examples 2.21 for Weibull and Gaussian Monte Carlo simulation

Solution	$\bar{b}$ (in)	$b_{\max}$ (in)	$\bar{b}$ (in)	$h_{\max}$ (in)	N
Weibull	0.7015	0.7196	1.4030	1.4391	2.36
Gaussian	0.875	0.8975	1.750	1.795	3.86

where $\bar{S} = 5788$ psi Eq. (2.143) and $\bar{s}$ is Eq. (2.133) with

$$\bar{s} = 1.06066 \frac{\bar{K}_t \bar{P} \bar{L}}{\bar{b}^3} \tag{2.157}$$

$$\bar{s} = 1,496$$

Substituting into Eq. (2.156) and Eq. (2.143)

$$N = \frac{5788}{1496} \tag{2.158}$$

$$N = 3.86$$

EXAMPLE 2.22. The same tip loaded cantilever beam used in Example 2.21 is used but the material is ductile Ti-16V-2.5Al titanium with the ultimate strength found in Example 1.5. The values for the conservative sizing of $\bar{b}$ for the beam is

Weibull

Eq. (1.81) low value for $\beta = 4.25$ (2.159)

Eq. (1.82) high value for conservative $\bar{b}$ $\Theta = 36.0853$ (2.160)

Eq. (1.83) low value for $\gamma = 141.000$ kpsi   (2.161)

Gaussian

Eq. (1.84) low value for $\mu = 176.684$ kpsi   (2.162)

Eq. (1.85) larger value for $\check{z} = 7.494$ kpsi   (2.163)

The goodness of fit shows these curves close, hence the $\bar{b}$ solutions should be closer than Example 2.21. Since a ductile material is being used, the stress will be different than the casting in Example 2.21. In Eq. (2.132) the K_t is used on the σ_a stress alone yielding for the titanium

$$s = [\sigma_a^2 + \sigma_m^2]^{1/2}$$

$$s = \left[\left(K_t \frac{\sigma_{max}}{2} \right)^2 + \left(\frac{\sigma_{max}}{2} \right)^2 \right]^{1/2} \tag{2.164}$$

$$s = \frac{\sigma_{max}}{2} [K_t^2 + 1]^{1/2}$$

Table 2.13 Cross sectional dimensions for Example 2.22 for Weibull and Gaussian simulation

Solution	$\bar{b}$ (in)	b_{max} (in)	$\bar{h}$ (in)	h_{max} (in)	N
Weibull	0.3279	0.3363	0.6558	0.6727	2.67
Gaussian	0.3813	0.3911	0.7626	0.7822	3.60

with $h = 2b$

$$s = \frac{6P_{max}L}{2\bar{b}(2\bar{b})^2}[K_t^2 + 1]^{1/2}$$

$$s = 0.75\frac{\bar{P}\bar{L}}{\bar{b}^3}[K_t^2 + 1]^{1/2}$$

(2.165)

with

$$\frac{\check{z}_{K_t}}{\bar{K}_t} = \pm 0.01163 \quad \frac{\check{z}_{\bar{p}}}{\bar{p}} = \pm 0.01 \quad \frac{\check{z}_{\bar{L}}}{\bar{L}} = \pm 0.01 \quad \frac{\check{z}_{\bar{b}}}{\bar{b}} = \pm 0.01$$

$$\bar{K}_t = 2.10 \quad \bar{p} = 30 \text{ lbs} \quad \bar{L} = 15'' \quad \bar{b} = \text{unknown}$$

Equation (2.165) is the Gaussian formulation to use with the Weibull and Gaussian curves for the titanium in separate Monte Carlo simulations to solve for $\bar{b}$ These are shown in Table 2.13 for $\bar{b}$, b_{max}, , h_{max}, and the safety factors. The material representation by Gaussian and Weibull curve are developed next.

Weibull

The development follows that in Example 2.21 and Eq. (2.128) except γ_L is 141,000 and from Fig. 2.26

$$\sigma_e' = \frac{S_{ult}}{2} = \frac{\gamma_L}{2}$$

(2.166)

$$\check{z}_{\sigma_e'} = \frac{S_{ult}}{30} = \frac{\gamma_L}{30}$$

(2.167)

$$C_{v\sigma_e'} = \frac{\gamma_L}{30}\frac{2}{\gamma_L} = \pm\frac{1}{15} = \pm 0.0667$$

(2.168)

and k_m is for 1×10^6 cycles and the values are the same now substitute into Eqs. (2.128) and (2.129)

$$\bar{\sigma}_e = \bar{k}_a\bar{k}_b \cdots k_m\sigma_e'$$

$$\bar{\sigma}_e = (0.8)(0.85)(0.88)\frac{141,000}{2} \text{ psi}$$

$$\bar{\sigma}_e = 42,187 \text{ psi}$$

(2.169)

Substituting into Eq. (2.130)

$$\check{z}_{\sigma e} = \bar{\gamma}_e'[(0.0667)^2 + (0.11)^2 + (0.06)^2 + (0.0455)^2]^{1/2}$$

$$\check{z}_{\sigma e} = \pm 0.1491\bar{\sigma}_e'$$

(2.170)

Note Eq. (2.168) $C_{v\sigma_e'} = \pm 0.0667$ and $C_{v\sigma_e} = \pm 0.1491$. Now using Eq.

(2.131)

$$(\sigma_e)_L = \gamma_{eL} = \bar{\sigma}_e[1 - 2.576(0.1491)]$$
$$\gamma_{eL} = 25,984 \text{ psi} \tag{2.171}$$

Now similar to Fig. 2.42 using γ_{line} as a failure line in Eq. (2.134).

$$\frac{1}{N} = \frac{\sigma_m}{\gamma_L} + \frac{\sigma_a}{\gamma_{eL}} \tag{2.172}$$

with

$$N = 1 \qquad \sigma_a/\sigma_m = 2.10$$

then

$$\sigma_a = 2.10 \; \sigma_m$$

substituting

$$1 = \frac{\sigma_m}{141,000} + \frac{2.10 \, \sigma_m}{25,984}$$
$$\sigma_m = 11,375 \text{ psi}$$
$$\sigma_a = 2.10(11,375) = 23,888 \text{ psi}$$

Now γ_S

$$\gamma_S = [\sigma_a^2 + \sigma_m^2]^{1/2}$$
$$\gamma_S = 26,458 \text{ psi} \tag{2.173}$$

minimum from Eq. (2.159)

$$\beta = 4.25 \tag{2.174}$$

minimum from Eq. (2.160)

$$\Theta = 36.085 \text{ kpsi} \tag{2.175}$$

The Weibull parameters Eqs. (2.172), (2.173), (2.174) are used with Eq. (2.165) in a Monte Carlo simulation to find $\bar{b}, b_{max}, \bar{h}, h_{max}$ and the safety factor. The next step is the Gaussian formulation for the material

Gaussian

     The failure line is the low side of a Gaussian curve in the same fashion as Example 2.21. On the σ_m axis Eq. (2.138) using Eqs. (2.162) and (2.163).

$$S_L = \mu_{min} - 2.576 \, \breve{z}_{max}$$
$$= 176,684 \text{ psi} - 2.576(7,494 \text{ psi}) \tag{2.176}$$
$$S_L = 157,379 \text{ psi}$$

The other end of the line is on σ_r axis and is $(\bar{\sigma}_e)_L$ as per Eqs. (2.139) and

(2.140). First using Eq. (2.128) and (2.129). But

$$\sigma'_e = \frac{S_{ult}}{2} = \frac{\mu_{min}}{2} = \frac{176,684 \text{ psi}}{2} \tag{2.177}$$

$$\sigma'_e = (0.8)(0.85)(0.88)\frac{176,684}{2}$$
$$\bar{\sigma}_e = 52,864 \text{ psi} \tag{2.178}$$

Now using the mean lines for $\mu = 176,684$ and $\bar{\sigma}_e = 52,864$ psi in Eq. (2.139)

$$\frac{1}{N} = \left[\frac{\sigma_m}{176,684} + \frac{\sigma_a}{52,864}\right]$$
$$\frac{\sigma_a}{\sigma_m} = 2.10 \quad \text{and} \quad N = 1 \tag{2.179}$$
$$1 = \left[\frac{\sigma_m}{176,684} + \frac{2.10\sigma_m}{52,864}\right]$$

$$\sigma_m = 22,034 \text{ psi}$$
$$\sigma_a = 2.10(22,034 \text{ psi}) = 46,271 \text{ psi} \tag{2.180}$$
$$\bar{\sigma}_S = [\sigma_a^2 + \sigma_m^2]^{1/2} = 51,250 \text{ psi}$$

Now Eq. (2.128) and Eq. (2.168)

$$\breve{z}_{\sigma e} = \bar{\sigma}_e[(0.0667)^2 + (0.11)^2 + (0.06)^2 + (0.0455)^2])^{1/2}$$
$$\breve{z}_{\sigma e} = \pm 0.1491 \, \bar{\sigma}_e \tag{2.181}$$

Now using Eq. (2.140)

$$\frac{1}{N} = \left[\frac{\sigma_{mL}}{S_L} + \frac{\sigma_{aL}}{(\sigma_e)_L}\right]$$
$$\bar{S}_L = 157,379 \text{ psi Eq. 2.176}$$
$$(\sigma_e)_2 = \bar{\sigma}_e - 2.576(0.1491\bar{\sigma}_e)$$
$$= 52,864[1 - 2.576(0.1491)]$$
$$(\sigma_e)_L = 32,560 \text{ psi} \tag{2.182}$$
$$\frac{\sigma_{aL}}{\sigma_{mL}} = 2.10 \, N = 1$$

Substituting

$$1 = \left[\frac{\sigma_{mL}}{157,379} + \frac{2.10\sigma_{mL}}{32,560}\right]$$

$$\sigma_{mL} = 14,114 \text{ psi} \tag{2.183}$$

$$\sigma_{aL} = 2.10(14,114 \text{ psi}) = 29,640$$

$$\sigma_{SL} = [\sigma_{mL}^2 + \sigma_{aL}^2]^{1/2} = 32,829 \text{ psi}$$

Now to obtain $\check{z}_S$ Eq. (2.141) using Eqs. (2.180) and (2.183)

$$2.576\check{z}_S = \bar{\sigma}_S - \bar{\sigma}_{SL}$$

$$= 51,250 - 32,829 \tag{2.184}$$

$$\check{z}_S = 7,151 \text{ psi}$$

Now

$$C_{VS} = \frac{\bar{\sigma}_S}{\check{z}_S} = \frac{7151 \text{ psi}}{51,250 \text{ psi}} = \pm 0.1395 \tag{2.185}$$

$$\bar{\sigma} = \bar{S} = 51,250 \text{ psi} \tag{2.186}$$

$$\check{z}_S = 7151 \text{ psi} \tag{2.187}$$

Equations (2.186) and (2.187) are used with Eq. (2.165) in a Monte Carlo simulation to find b, b_{max}, h, h_{max}, and safety factor from Weibull Eqs. (2.144)–(2.151) with $\bar{b} = 0.3279$ and Gaussian Eqs. (2.152)–(2.158) with $\bar{b} = 0.3813$ in Table 2.13. All iterations started from the low side.

G. Bounds on Monte Carlo Fatigue Calculations

1. The minimum P_f for a structural member stress s_1.
There are some features of a Monte Carlo or probability solution which should be discussed. They are, how small can the P_f be, and find a factor of safety, N. How to select t and hence the P_f to produce N. First the C_{vS} and C_{vs} for material and the stress due to loading should be developed.
The value for C_{vS} is from Eqs. (2.186) and (2.187)

$$C_{vS} = \frac{\check{z}_S}{S}$$

$$C_{vS} = \frac{7151}{51,250} = \pm 0.1395 \tag{2.188}$$

Now to develop C_{vs} using a card sort for $\bar{s}$ Eq. (2.165)

$$\bar{s} = 0.75\frac{\bar{P}\bar{L}}{\bar{b}^3}[\bar{K}_t^2 + 1]^{1/2} \tag{2.189}$$

K_t will be used as [2.10].

$$\bar{s} = 0.75\frac{30\,\text{lb}\,(15\,\text{in})}{\bar{b}^3}[(2.1)^2 + 1]^{1/2} = \frac{785.005}{\bar{b}^3} \tag{2.190}$$

Now select maximum values for P, L, K_t and minimum values for b

$$L_{\max} = L[1 + 2.576(0.01)] = 15.3864 \text{ in}$$
$$b_{\min} = \bar{b}[1 - 0.02576] = 0.97424\,\bar{b}$$
$$P_{\max} = \bar{P}[1.02576] = 30.7728 \text{ lbs}$$
$$K_{t\max} = \bar{K}_t[1.02576] = 2.1541$$

Note four cards or parameters are selected $\bar{x} = 6.0737$ Table 2.2 and Eq. (2.25)

$$\bar{x}\,\check{z}_s = s_{\max} - \bar{s} \tag{2.191}$$

$$s_{\max} = 0.75\frac{(30.7728)(15.3864)}{(0.97424\bar{b})^3}[(2.1541)^2 + 1]^{1/2}$$

$$s_{\max} = \frac{912.037}{\bar{b}^3} \tag{2.192}$$

Substituting Eq. (2.190) and (2.192) into Eq. (2.191) Now

$$6.0737\,\check{z}_s = \frac{912.037}{\bar{b}^3} - \frac{785.005}{\bar{b}^3}$$

$$\check{z}_s = \frac{20.9152}{\bar{b}^3} \tag{2.193}$$

Now

$$C_{vs} = \frac{\check{z}_s}{s}$$

$$C_{vs} = \frac{20.9152}{\bar{b}^3}\frac{\bar{b}^3}{785.005} - \pm 0.0266433 \tag{2.194}$$

Now using the coupling Eq. (2.42)

$$t = -\frac{\mu_S - \mu_s}{[\check{z}_S^2 + \check{z}_s^2]^{1/2}} \tag{2.195}$$

Using Eq. (2.186)

$$\mu_S = \bar{S} = 51{,}250 \text{ psi}$$
$$\mu_s = \bar{s} \text{ which is to be solved for}$$
$$\check{z}_S = C_{vS}\bar{S} \text{ Eq. (2.188)}$$

From Eq. (2.194)

$$\check{z}_s = C_{vs}\bar{s}$$

Now substituting in to the coupling Eq. (2.195) and squaring

$$t^2 = \frac{(\bar{S} - \bar{s})^2}{(C_{vS}\bar{S})^2 + (C_{vs}\bar{s})^2} \tag{2.196}$$

Solving for an unknown $\bar{s}$

$$t^2[C_{vS}^2\bar{S}^2 + C_{vs}^2\bar{s}^2] = \bar{S} - 2\bar{S}\bar{s} + \bar{s}^2$$
$$(1 - t^2C_{vs}^2)\bar{s}^2 - 2\bar{S}\bar{s} + (\bar{S}^2 - t^2C_{vS}^2\bar{S}^2) = 0$$
$$A\bar{s}^2 + B\bar{s} + C = 0$$

Now solving the quadratic equation

$$\bar{s} = \frac{-B \pm \sqrt{B^2 - 4AC}}{2A} \tag{2.197}$$

Now examine the solutions for S Eq. (2.197)

1. If $A = 0$ then s is infinite and the stress due to loading is greater than $\bar{S}$ which is the stress the material can resist (Example 2.14) hence a failure.

$$A = 1 - t^2C_{vs}^2$$
$$\left[-t^2 = -\frac{1}{C_{vs}^2}\right] \tag{2.198}$$

   substitute Eq. (2.194)

$$t = \frac{1}{C_{vs}} = \frac{1}{\pm 0.0266433}$$
$$t = 37.53$$

2. If $C = 0$ then

$$C = \bar{S}^2 - t^2C_{vS}^2\bar{S}^2$$

   then $\bar{S}^2 \neq 0$ but

$$1 - t^2 C_{vS}^2 = 0$$

$$t = \frac{1}{C_{vS}}$$

Noting Eq. (2.42) and substituting Eq. (2.188)

$$t = \frac{1}{\pm 0.1395} = \pm 7.168$$

Table 2.3

$$P_f \approx 10^{-12}$$

Also note Eq. (2.188) inverted

$$t = \frac{\bar{S}}{\bar{z}_S} = \frac{51{,}250}{7151} = 7.168 \tag{2.199}$$

Now substitute $C = 0$ into Eq. (2.197)

$$\bar{s} = \frac{-B \pm B}{2A} \tag{2.200}$$

for $+B$

$$\bar{s}_1 = 0$$

This is a structural member which cannot be sized since member size approaches infinity.

for $-B$

$$\bar{s}_2 = \frac{-(-B) - (-B)}{2A}$$

$$\bar{s}_2 = \frac{4(\bar{S})}{2(1 - t^2 C_{vs}^2)} \tag{2.201}$$

$$\bar{s}_2 = 2.0757\bar{S}$$

$\bar{s}_2$ is a safety device not a structural stress like $\bar{s}_1$. It would appear that t greater than Eq. (2.199) does not allow a structural design with the Gaussian–Gaussian formulation for a structural member. Note the Gaussian–Weibull formulation may not be as severely limited as γ is the lowest stress in the Weibull formulation.

    2. t and P_f in terms of the safety factor N

        Now examine Eqs. (2.42) and (2.196)

$$t^2 = \frac{(\bar{S} - \bar{s})^2}{(C_{vS}\bar{S})^2 + (C_{vs}\bar{s})^2}$$

Divide top and bottom by s and set $N = \bar{S}/\bar{s}$ for structural factor of safety

$$t^2 = \frac{(N-1)^2}{(C_{vS}N)^2 + C_{vs}^2} \tag{2.202}$$

from Eq. (2.194) $C_{vs}^2 = (0.0266433)^2 = 0.000709$
If $N = 1$ $C_{vS} = \pm 0.1395$ Eq. (2.188) then

$$(C_{vS}N)^2 = 0.01946$$

In this case dropping C_{vs}^2 causes in the denominator the following percent error for $N = 1$

$$\left[1 - \frac{C_{vS}^2}{(C_{vS}N)^2 + (C_{vs})^2}\right] \times 100 = 3.52\%$$

Now drop C_{vs}^2 in Eq. (2.202) and take the square root

$$t = \frac{N-1}{C_{vS}N} \tag{2.203}$$

Now with $C_{vS} = \pm 0.1395$ Eq. (2.188)

$$t = \pm 7.168 \frac{N-1}{N} \tag{2.204}$$

If a safety factor of 3 is desired

$$t = \frac{3-1}{3}(\pm 7.168)$$

$$t = \pm 4.7787$$

Table 2.3

$$P_f \cong 10^{-6}$$

Now when $t = \pm 7.168$ in Eq. (2.199) substituting in Eq. (2.204)

$$\pm 7.168 = \pm 7.168 \frac{N-1}{N}$$

$$\pm N = \pm[N-1]$$

for $+$ sign

$$N = N - 1$$

$$0 = -1$$

A contradiction and for $-$ sign

$$-N = -[N - 1]$$
$$0 = +1$$

Another contradiction hence no structural factor of safety for the Gaussian–Gaussian formulation. The Gaussian–Weibull may be again not be as severely limited. The conclusions and useful tools are shown in Table 2.14 including calculation limitations for Gaussian distributions

H. Approximate Dimension Solution Using Cardsort and Lower Material Bounds

Now attempt to formulate a means to use an upper bound on a cardsort for s, the stress due to loads and set it equal to a lower bound on the material properties. A cantilever beam sized in fatigue in Example 2.22 is used with the $\sigma_a - \sigma_m$ curve developing S for ductile titanium Ti-16V-2.5Al Eqs. (2.173)–(2.175) for the Weibull distribution and Eqs. (2.186) and (2.187) for the Gaussian distribution. The original tensile strength data is Example 1.5 for 755 samples which fits both Weibull and Gaussian curves equally well.

The Gaussian form of the stress due to the load, s, for a maximum possible s_{max} is 6.0737 standard deviations above the mean, $\bar{s}$, with a probability of not being exceeded of approximately $1/10^9$. This is developed in Eqs. (2.190)–(2.192)

$$s_{max} = \frac{912.037}{\bar{b}^3}$$

$$P(s_{max} >) = \frac{1}{10^9} \tag{2.205}$$

The s_{max} will be set equal to a lower value of the Gaussian and the Weibull distributions.

Table 2.14 Calculations limitations for Gaussian distributions

Gaussian–Gaussian	Gaussian–Weibull
A. $t \leq \dfrac{\bar{S}}{z_S}$ Eq. (2.199)	A. t not as limited here
with $P_f \geq 10^{-12}$ Table 2.3	
B. $t \leq \dfrac{N-1}{C_{vS} N}$ Eq. (2.203)	B. again t not as limited

Gaussian–Gaussian

The material representation is Eqs. (2.186) and (2.187)

$\bar{S} = 51,250$ psi

$\check{z}_S = 7151$ psi

A lower value K_c 99.999% greater with 95% confidence from Appendix E where $\alpha_c = \bar{x} - K_c S$ with $K_c = 4.8$ from Fig. E1 for 100 samples.

$\alpha_c = 51,250$ psi $- 4.8(7151$ psi$)$

yields

$$\alpha_c = 16,925 \text{ psi} \tag{2.206}$$

Now set s_{max} Eq. (2.205) equal to α_c Eq. (2.206)

$$\bar{b}^3 = \frac{912.037}{16,925} \tag{2.207}$$

$\bar{b} = 0.3777$ in (0.3813 Example 2.22; Table 2.13)

from Eq. (2.3777)

$\bar{b}_{max} = (1 + 0.02576)\bar{b}$

$$\bar{b}_{max} = 0.3874 \text{ in } (0.3911 \text{ Example 2.22; Table 2.13}) \tag{2.208}$$

and

$h = 2b$

Now to calculate $\bar{b}$ for

Gaussian–Weibull

The Weibull representation of the material is Eqs. (2.173)–(2.175) with $\gamma_s = 26,458$ psi Eq. (2.173) and setting it equal to Eq. (2.205)

$$\bar{b}^3 = \frac{912.037}{26,458} \tag{2.209}$$

$$\bar{b} = 0.3255 \text{ in } (0.3279 \text{ Example 2.22; Table 2.13}) \tag{2.210}$$

$$\bar{b}_{max} = 0.3338 \tag{2.211}$$

Probability of failure and safety factor

Gaussian–Gaussian

The overlap areas of stress and material will give the total failure P_f. The P_f material is $1/10^5$ and $P_f(s_{max} >) = 1/10^9$

$$P_f = \frac{1}{10^5} + \frac{1}{10^9} = \frac{1}{10^5} \tag{2.212}$$

The factor of safety

$$N = \frac{\bar{S}}{\bar{s}} \tag{2.213}$$

with $\bar{s}$ Eq. (2.190)

$$\bar{s} = \frac{785.005}{\bar{b}^3} \tag{2.214}$$

and Eq. (2.207)

$$\bar{s} = 14,569 \text{ psi} \tag{2.215}$$

and substituting into Eq. (2.213)

$$N = \frac{51,250}{14,569} \tag{2.216}$$
$$N = 3.52$$

Gaussian–Weibull
The probability of failure P_f is that of $P(s_{max} >) = \frac{1}{10^9}$ since $\gamma_s = 26,458$ psi is the lowest value of the material representation

$$P_f = P(s_{max} >) = \frac{1}{10^9} \tag{2.217}$$

Now to find S(50 percentile) for the Weibull representation. From Eq. (2.149) set up

$$\frac{x - \gamma}{\Theta} = 0.693147^{1/\beta} \tag{2.218}$$

x is the 50 percentile value so substituting Eqs. (2.173)–(2.175)

$$\frac{x - 26,458}{36,085} = (0.693147)^{1/4.25}$$
$$\bar{S} = 59,562 \text{ psi}$$

Substitute $\bar{S}$ into Eq. (2.213) and Eq. (2.214) with b from Eq. (2.210) into the factor of safety.

$$N = \frac{59,562}{22,763} \tag{2.219}$$
$$N = 2.617$$

These approximate values compare closely with values in Table 2.13 which is a Monte-Carlo simulation with $P_f = 1/10^6$.

REFERENCES

2.1. Aerospace Structural Metals Handbooks CINDAS/USAF CRDA. Purdue University, West Lafayette, Indiana.

2.2. AGMA 218.01, Pitting Resistance and Bending Strength of Spur and Helical Gears, Arlington, AGMA, 1982.

2.3. Boller CHR, Seegar T. Material Data for cyclic Loading, Elsevier, 1987.

2.4. Castleberry G. Mach. Des. 50(4):108–110, February 23, 1978.

2.5. Dieter GE. Mechanical Metallurgy 3rd Ed, New York: McGraw-Hill Publishing Co, 1986.

2.6. Dieter GE. Engineering Design, New York: McGraw-Hill Book Company, 1983.

2.7. Dixon JR. Design Engineering, New York: McGraw-Hill Book Company, 1966.

2.8. Deutschman AD, Michaels WJ, Wilson CE. Machine Design, New York: MacMillan Publishing Co. 1975.

2.9. Faires VM. Design of Machine Elements and Problem Book, New York: The MacMillan Co, 1965.

2.10. Faupel JH, Fisher FE. Engineering Design, New York: John Wiley and Sons Inc, 1981.

2.11. Forrest PG. Fatigue of Metals, Reading, Mass: Addison-Wesley, 1962.

2.12. Frost, NE, Marsh KJ, Pook LP. Metal Fatigue, London: Oxford University Press, 1974.

2.13. Fry TR. Engineering Uses of Probability, D. Van Nostrand, 1965.

2.14. Good IS. Probability, Hafner, 1950.

2.15. Grover HJ, Gordon SA, Jackson LR. Fatigue of Metals and Structures, NAVWEPS Report 00-25-534, Bureau of Naval Weapons, Department of the Navy, Washington, D.C., 1960.

2.16. Hagendorf, HC, Pall FA. A Rational Theory of Fatigue Crack Growth, NA-74-278, Rockwell International, Los Angeles, CA, (1974).

2.17. Haugen EB. Probabilistic Approaches to Design, University of Arizona, Summer Course, Arizona, 1971.

2.18. Haugen EB. Probabilistic Mechanical Design. New York: Wiley Science, 1980.

2.19. Haugen EB, Wirsching PH. Probabilistic Design Reprints Machine Design 17 April 25–12 June 1975, Cleveland, Penton, Inc, 1975.

2.20. Hine, CR, Machine Tools and Processes for Engineers, New York: McGraw-Hill Book Co. 1971.

2.21. Hodge JL, Lehmann EL. Elements of Finite Probability, Holden-Day, 1970.

2.22. Horowitz J. Critical Path Scheduling, New York: Ronald Press Co, 1967.

2.23. Johnson NL, Leone FC. Statistics and Experimental Design New York: John Wiley and Sons, 1964.

2.24. Johnson RC. Optimum Design of Mechanical Elements, New York: John Wiley, 1961.

2.25. Johnson RC. Mach. Des., 45(11):108, May 3, 1973.

2.26. Juvinall RC. Stress, Strain, and Strength, New York: McGraw-Hill Book Co, 1967.

2.27. Kececioglu DB, Chester, LB. Trans, Soc. Mech. Eng. (J. Eng. Ind.), 98(1); Series B:153–160, February 1976.

2.28. Kemeny JG, Snell JL, Thompson GL. Introduction to Finite Mathematics, Englewood Cliffs: Prentice-Hall, 1957.

2.29. Kliger HS. Plast. Des. Forum, 2(3):36–40, May/June 1977.

2.30. Landau D. Fatigue of Metals—Some Facts for the Designing Engineer, 2nd ed., New York: The Notralloy Corp, 1942.

2.31. Lindley DV. Making Decision, New York: Wiley Interscience, 1971.

2.32. Lipson C, Juvinall RC. Stress and Strength, New York: MacMillan Co, 1963.

2.33. Lipshultz S. Finite Mathematics, New York: McGraw-Hill Schaums Outline, 1966.

2.34. McMaster RC. Non-Destructive Testing Handbook, Vol. I, New York: Ronald Press, 1959.

2.35. Manson SS. (1965) Experimental Mechanics, 5(7):193, 1965.

2.36. Metals Hdbk. Supplement, Cleveland, The American Society for Metals, 1954.

2.37. Meyer P. Introduction to Probability and Statistics Application, Addison Wesley, 1965.

2.38. Miller I, Freund JE. Probability and Statistics for Engineers, Englewood Cliffs, NJ: Prentice-Hall, 1965.

2.39. Middendorf WH. Engineering Design, Boston: Allyn and Bacon Inc, 1969.

2.40. Miner DF, Seastone JB. Handbook of Engineering Materials, New York: John Wiley and Sons Inc, 1955.

2.41. Mischke CR. ASME Paper 69-WA/DE-6 A method relating factor of safety and reliability. ASME Winter Annual Meeting 1969, Los Angeles, CA 1960.

2.42. Mischke CR. Rationale for Design to a Reliability Specification, New York: ASME Design Technology Transfer Conference, 5–9 Oct. 1974.

2.43. Mischke CR. Winter Annual Meeting 1986 ASME, Anaheim, CA, 1986. ASME Paper 86-WA/DE-9 A New Approach for the Identification of a Regression Locus for Estimating CDF-Failure Equations on Rectified Plots. ASME Paper 86-WA/DE-10 Prediction of Stochastic Endurance Limit. ASME Paper 86/DE-22 Some Guidance of Relating Factor of Safety to Risk of Failure. ASME Paper 86/DE-23 Probabilistic Views of the Palmgren - Minor Damage Rule.

2.44. Miske CR. Stochastic Methods in Mechanical Design Part 1: Property Data and Weibull Parameters. Part 2: Fitting the Weibull Distribution to the Data. Part 3: A Methodology. Part 4: Applications Proceedings of the Eighth Bi-Annual Conference on Failure Prevention and a Reliability, Design Engineering D.V. of ASME, Montreal, Canada, Sept. 1989. To be published in Journal of Vibrations, Stress and Reliability in Design, 1989.

2.45. Morrison JLM, Crossland B, Parry JSC. Proc. Inst. Mech. Eng. (London), 174(2):95–117, 1960.

2.46. Mosteller F, Rourke REK, Thomas IR, GB. Probability with Statistical Applications, Addision Wesley, 1961.

2.47. Osgood CC. Fatigue Design, New York: Wiley-Interscience, 1970.

2.48. Owen MJ. Fatigue of Carbon-Fiber-Reinforced Plastics, In: Broutman LJ, Krock RH eds, Composite Materials, Vol. 5, New York: Academic Press, 1974.

2.49. Peterson RE. Stress Concentration Design Factors, New York: John Wiley and Sons, 1974.

2.50. Salkind MJ. Fatigue of composites, Composite Materials, STP 497, ASTM, Philadelphia, PA, 1971.

2.51. Shigley JE, Mischke CR. Mechanical Engineering Design, New York: McGraw-Hill Book Co, 1989.

2.52. Sines G, Waisman J. eds, Metal Fatigue McGraw-Hill Book Company, 1959.

2.53. Siu WWC, Parimi SR, Lind NC. Practical Approach to Code Calibration J. Structural Division ASCE, July 1975.

2.54. Sors L. Fatigue design of machine components, Pergamon Press, Oxford, 1971.

2.55. Tribus M. Rational Description, Decision, and Designs, Pergamon Press, 1969.

2.56. Wirsching PH, Kempert JE. Mach. Des., 48(21):108–113, September 23, 1976.

2.57. Von Mises R. Mathematical Theory of Probability and Statistics, Academic, 1964.

2.58. Von Mises R. Probability Statistics and Truth, 2nd ed, New York: MacMillan, 1957.

2.59. Weihsmann P. Fatigue Curves with Testing, New York, M.E. March 1980.

2.60. An Index of US Voluntary Engineering Standards, Slattery WJ, ed., NBS 329 plus Supplements 1 and 2, US Government Printing Office, Washington, D.C., 1971.

2.61. ASME Boiler and Pressure Vessel Code, The American Society of Mechanical Engineers, United Engineering Center, 345 E. 47th St. New York, N.Y., 1977.

2.62. Machinery's Handbook, 20th ed. New York; Industrial Press Inc. 1975.

2.63. Metals Handbook Vol. I–V, 8th ed., American Society for Metals, Metals Park, OH.

2.64. Steel Construction, 7th ed., AISC Manual, American Institute of Steel Construction, 101 Park Ave, New York, NY, 1970.

2.65. Strength of Metal Aircraft Elements, Military Handbook MIL-HDBK-5F. Washington, DC, 1990.

2.66. Timber Construction Manual, 2nd ed. AITC, New York: John Wiley and Sons Inc, 1974.

2.67. Smith R, Hirshberg M, Manson SS. Fatigue Behavior of Materials Under Strain in Low and Intermediate; NASA Technical Note No. D1574.

PROBLEMS

PROBLEM 2.1

A rectangular cross-section beam Fig. Prob. 2.1 is to be used to support a chain hoist. Neglect the weight of the beam.

$$S = \frac{MC}{I}$$

If $I = bh^3/12$ and coefficients of variations are

$$C_L = \pm 2\%; \quad C_S = \pm 10\%; \quad C_b = C_h = \pm 1\%$$

Find: $\left(\dfrac{\check{z}_p}{\bar{p}}\right)$ with $\bar{P} = 1000$ lbs and F.S. $= 1.25$

PROBLEM 2.2

Assuming variations $C_S = \pm 5.5\%$ and $C_p = \pm 5.\%$ with F.S. $= 1.25$ in Problem 2.1 and C_b, and C_h are $\pm 1\%$.
    Compute $\bar{L}$. Compute the P_f

PROBLEM 2.3

In designing spherical fuel tanks for a rocket engine, the internal diameters are 10.00 ± 0.05 in $(P = 0.99)$ and the specific weight of the fuel is 50.0 ± 0.6 lbm/cu.ft. $(P = 0.99)$. Show which variable contributes the great-

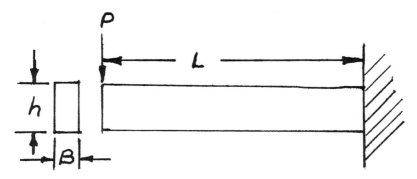

Figure Problem 2.1

est percent uncertainty in computing estimates of the weight of the contents of the tanks.

PROBLEM 2.4

Flow in a circular pipe (laminar flow) is given by

$$Q = \frac{\pi d^4 \gamma h_L}{128 \mu L}$$

What is the percentage uncertainty in the flow rate if:

$C_d = 1\%$	$P = 0.99$	$C_\gamma = 3\%$	$P = 0.99$
$C_{hL} = 2\%$	$P = 0.99$	$C_\mu = 3\%$	$P = 0.99$
$C_L = 3\%$	$P = 0.99$		

Which uncertainty contributes most to the uncertainty in Q and find C_Q.

PROBLEM 2.5

The following data represent the pumping ability of a certain type of pump. (i.e. ten pumps of the same kind.)

OUTPUT (gal/min)	101	103	90	97	98	100	100	102	99	110
Pump#		1	2	3	4	5	6	7	8	9	10

The following table represents the pumping requirements for a country estate:

DAY	Mon	Tues	Wed	Thurs	Fri	Sat	Sun
req.gpm	100	70	70	90	80	90	60

Assume that the above data are samples from Gaussian distributions. When using a pump of the type given above, what is the probability of failure? (i.e. not enough capacity.)

What is the factor of safety of the pump-country estate combination?

PROBLEM 2.6

An equilateral triangle, height $b = 25.62'' \pm 0.010''$, is hung from long wires and used as a torsional pendulum oscillating about a vertical axis. The test sample center of gravity is above that of the plate and the test sample moment of inertia is determined about a vertical line. The I_T from *Weights*

Engineering Handbook, Society of Allied Weight Engineers, Los Angeles (1976) is

$$I_T = I_p \left[\frac{(W_T + W_p)}{W_p} \left(\frac{T_{T+P}}{T_P} \right)^2 - 1 \right]$$

where $I_P = \dfrac{W_p b^2}{12}$

W_T = Weight Test Sample

W_P = Weight of Plate

$T_{(T+P)}$ – Period of vibration plate plus test sample

T_P – Period of vibration plate alone

Find C_V and 2.576 $\check{z}_{IT}$ by a cart sort solution when

$$W_p = 14.343 \text{ lbs} \quad W_T = 29.625a \text{ lbs} \quad T_p = 1.1725 \text{ sec/cycle}$$
$$T_{T+P} = 1.665 \text{ sec/cycle}$$

and

weights to ±0.01 grams and time to ±0.01 sec

PROBLEM 2.7

A shear washer in Fig. Prob. 2.7 is to fail as a safety device through the thickness, with an applied force 990 lbs ± 10 lbs. Use

$$\tau_{ult} = \frac{P}{\pi d t}$$

with

$d = 0.255 \pm 0.005$

$t = t \pm 0.001$

$P = 990 \text{ lbs} \pm 10 \text{ lbs}$

Using a bronze with σ_{ult} 80,000 psi find t using Eq. (2.42) with the left hand side t of ±4 for a safety device and compare with t for a structural member.

PROBLEM 2.8

A hardened steel pin Fig. Prob. 2.8 with 180,00 psi ultimate strength is pressed into a 6061-T6 aluminum flange. The pin diameter is

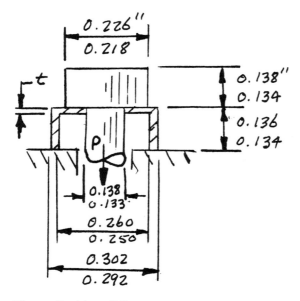

Figure Problem 2.7

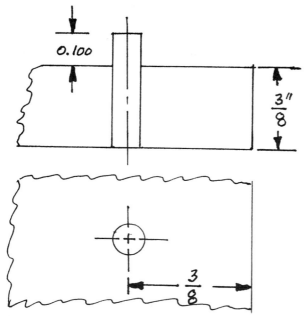

Figure Problem 2.8

0.0940–0.0935 in. while the hole is 0.0932–0.0930 inches. Find the mean and standard deviation for the combined stress near the pin in the flange, as well as the factor of safety and probability of failure for the 6061-T6 aluminum flange. Also, find the mean and standard deviation of the force to press the pin in.

Use the easier card sort solution.

PROBLEM 2.9

A gull wing solder tab Fig. Prob 2.9 is modeled as a beam (solder tab) resting on an elastic foundation (solder) which has below the tab a circuit board (assumed rigid). The forces are

$$F_x = F_y = F_z = 1 \text{ oz} \pm \frac{1}{3} \text{oz}$$

$$h = L = 2b = 0.100'' \pm 0.010''$$

$$E_{solder} = 4 \times 10^6 \text{ psi} \quad E_{copper} = 16 \times 10^6 \text{ psi}$$

$$t_c = 0.020'' \pm 0.002''$$

$$0.004'' \leq t \leq 0.010''$$

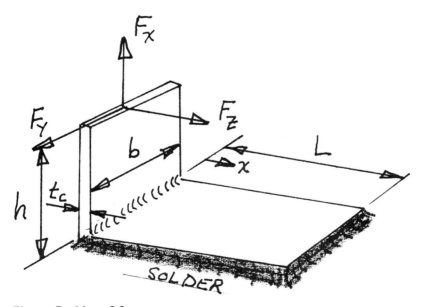

Figure Problem 2.9

From W. Griffel's, *Handbook of Formulas for Stress and Strain,* Fredrick Ungar Company (1966) and M. Hetenyi, *Beams on Elastic Foundations,* University of Michigan Press (1946)

1. Y_{mo} deflection at $x=0$ of $M_o = hF_z$

$$Y_{mo} = \frac{2M_o\lambda^2}{K}$$

where

$$\lambda = \left[\left(\frac{E_S}{E_c}\right)\frac{12}{t_S t_c^3}\right]^{1/4}$$

$$K = \frac{bE_S}{t_S}$$

2. Y_P deflection at $x=0$ of $P = F_x$

$$Y_P = \frac{2F_x\lambda}{K}$$

3. Y_{To} deflection at $x=0$ of $T_o = hF_y$

$$Y_{To} = \frac{b}{2}\phi_x$$

with

$$\phi_x = T_o\frac{12\alpha}{K_0 b^3}\frac{\cosh \alpha L}{\sin \alpha L}$$

where

$$K_0 = \frac{K}{b} = \frac{E_S}{t_S}$$

$$\alpha = \left[0.78\left(\frac{E_S}{E_c}\right)\frac{b^2 + t_c^2}{t_S t_c^3}\right]^{1/2}$$

4. The solder stress is

$$\sigma_S = E_S\frac{Y_{total}}{t_S} = E_S\frac{Y_{mo} + Y_P + Y_{To}}{t_S}$$

Find the mean and standard deviation of the stress using a card sort solution.

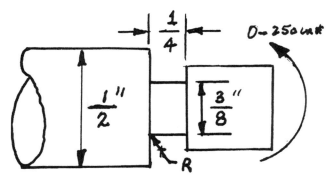

Figure Problem 2.10

PROBLEM 2.10

Using titanium Ti-16V-2.5 AL and using Eqs. (2.17) and (2.175) in with Eq. (2.173) modified for the stress concentration. The stationery shaft Fig. Prob. 2.10 has an applied moment of 0–250 in lbs and the original R was 0.002 in. What would be the probability of failure for the Gaussian and Weibull representation of the titanium? What R should be used to produce a $P_f = 1/10^6$.

3
Optimum Design

I. FUNDAMENTALS

The concept behind optimum design is to get the most from an engineering design with the least cost, effort, or materials. The analysis generally starts with a

A. Criterion Function

$$C = C(x, \cdots z_n)$$

Generally there is only one such function and it can represent any or all of the following

1. Cost of manufacturing a product
2. Total weight of a design
3. Power developed by a design
4. Energy absorbed by a design
5. Efficiency of the design

These are some of the factors which the engineer must take into account to get the least or most out of their design with the criterion function. There are other constraints that must be satisfied:

B. Functional Constraints

These are the physical laws which an engineer must follow for a successful design

1. Stress equations
2. Calculations for thrust, lift, drag
3. Deflections
4. Buckling

Functional constraints are factors which engineers have studied in their undergraduate years and dealt with during career.

The last of the constraints is the regional constraint.

C. Regional Constraints

These are the physical limitations

1. Number of men in a shop
2. Gross weight of an aircraft not to exceed the design weight of the runway
3. Truck widths less than one lane of a highway

These generally state that a parameter must be less or greater than a stated number. The design criterion most engineers use are the equations taught in undergraduate course work which are called functional constraints. Then to fabricate the design, vendors are selected and the lowest cost or criterion function to fabricate is selected. The maximum sizes or limitations on the design are regional constraints which for the most part are common sense. However, when optimizing with a computer or employing an analyst all of the above criterion function, functional constraints, and regional constraints must be stated as clearly as possible. The first solution will show if the equations are not constrained by giving answers like infinity, zero, or a negative dimension (all real values greater than zero positive). Various industries optimize differently depending upon their specific concerns.

II. INDUSTRY OPTIMAL GOALS

A. Flight Vehicles

Light weight is the most important consideration, therefore, the stress is pushed as high as possible. It would appear that σ_{ult}, σ_{yt} would be the most important consideration. However, the thin wall cross sections will often buckle below the compression yield of the material and can fail by fatigue even when the material is in compression. The loads causing stress when properly defined are always varying. Variation in the tension stresses causes fatigue and development of cracks in the thin walls. As if this is not enough, thermal stresses will cause creep above certain temperatures. In turn the performance created by the engines must be prescribed to develop proper design parameters. The optimization entails

1. Criterion function
    Maximize performance
    Minimize weight
    Minimize cost (lastly)
2. Functional
    All the equations to properly define the vehicle function. This can be several computer programs, or analysis of structural vibrations and response which present stiffness requirements
3. Regional constraints
    Vehicle weight is a maximum value often dictated by their runway capacity
    Must fit in a prescribed space
    Crew size
    Travel a given speed
    Design life

B. Petro or Chemical Plants

These plants function for 30–50 years and are expected to survive major earthquakes and explosions with as little damage as possible. The criterion here is a process to produce a product and the design of pressure vessels, pipes, and physical containment of the parts of the process. When vessels are welded a proof test of 140% of the operating pressure is applied to cause a failure should a critical crack size exist. Corrosion allowance of 1/16 to 1/4 inch on one or both sides of a vessel are added to the design thickness. Thin wall vessels (which are optimal) are a minimum weight structure for the applied loads.

1. Criterion function
    Maximize profitable life
    Minimum cost
2. Function constraints
    All equations to define a process or vessel being designed
3. Regional constraint
    Weight and size of vessels or components being designed as they must be sent by trucks over roads, barged down waterways, or on railroad flatbeds over existing railroads. Bridge clearance and total land area are also problems here

C. Main and Auxiliary Power and Pump Units

The criterion here is high reliability and excellent performance. The strength of the parts, cavitation, and creep are prime considerations. Means of fabrication become critical in holding down costs. In power units the emissions due to burning of fossil fuel becomes a challenge and are not readily solved. Vibration becomes a problem.

1. Criterion function
   Maximize performance
   Minimize cost
2. Function constraints
   All equations describing the system
3. Regional constraints
   Volume and weight
   Emissions

D. Instruments and Optical Sights

The operating character of the main equipment (tanks, ships, helicopters...) is important and becomes a design parameter for the auxiliary equipment. Servo systems that control them have natural frequencies generating instabilities in the system. The design criterion is based on frequency and small deflections and rotations to minimize instrument and sight errors. Note: With this criterion the stress is seldom large and does not present a problem but is present mostly in the computation of frequency and optical errors.

1. Criterion function
   Maximize functional performance
   Minimize weight as unit must fit in a confined space
   Minimize cost but not sacrifice performance
2. Functional constraints
   Frequencies, deflections, and functional performance are critical
3. Regional constraints
   Sizes and weight
   Optical rotational and deflection limits

E. Buildings or Bridges

The design criterion requires that on unsupported spans, deflections are less than L/360 where "L" is in inches (keeps drywall, tile, etc. from cracking). Bridges are arched to minimize deflections (imagine watching a car or truck in front of you sinking relative to you on a bridge and what your reaction would be!). Buckling and wind loads affect building complexes and bridges. Earthquakes also are a major concern, such as the 1994 Northridge earthquake which cracked many welded joints in steel structures (the fixes for these are still being studied). The material behavior of the beam and columns are vital, in bridge construction as older chain link bridge material has changed with time, causing the bridge to fall into the river sometimes with people on the span. Structural members are minimized for cross sectional area and weight for the applied loads.

1. Criterion function
   Minimize cost
   Maximize life of structure
2. Functional constraints
   Buckling, tension stresses due to design loads
3. Regional constraints
   Size and height

F. Ships or Barges

Ships or barges are buoyed upward by displaced water that provides a uniform elastic support to the structure. The ships and barges roll and pitch overall as well as elastically deflect along the length. The members must withstand the fatigue stresses and buckling loads.

1. Criterion function
   Maximize cargo
   Minimize costs
2. Functional constraints
   Buckling and fatigue loads
3. Regional constraints
   Size to fit in the canals at Panama and Suez, berths at ports
   Draft to fit in most channels and harbors

III. OPTIMIZATION BY DIFFERENTIATION

The optimum design is obtained by several mathematical techniques. When all functional constraints can be substituted into a criterion function the derivative with respect to each variable may be set equal to zero. For n variables this can yield n equations for a solution.

EXAMPLE 3.1 [3.23]. An open top rectangular tank with its base twice as long as wide is to have a volume of 12 cubic feet. Determine the most economical dimensions, if the bottom sheet material cost 0.20 dollars per square foot and the sides 0.10 dollars per square foot.

Let: a – width of base $2a$ – length of tank
        b – height of tank

Dollar cost of the bottom

$$C_B = a(2a)(0.20) = 0.40a^2$$

Dollar cost of the sides

$$C_S = 2(ab)(0.10) + 2(2a)(b)(0.10) = 0.60\,ab$$

The total cost of criterion function

$$C = 0.40\,a^2 + 0.60\,ab$$

The functional constraint

$$V = (\text{base})\,(\text{width})\,(\text{height}) = 12\ \text{ft}^3$$
$$V = (2a)(a)(b) = 12\ \text{ft}^3$$
$$b = \frac{6}{a^2}$$

Now placing the functional constraint into the criterion function

$$C = 0.40\,a^2 + 0.60\,ab = 0.40\,a^2 + 0.60\,a\left(\frac{6}{a^2}\right)$$
$$C = 0.40\,a^2 + \frac{3.60}{a}$$

To obtain the minimum cost

$$\frac{dC}{da} = 0.80\,a - \frac{3.60}{a^2} = 0$$
$$a^3 = \frac{3.60}{0.80}$$
$$a = 1.65\ \text{ft} \quad b = \frac{6}{a^2} = 2.20\ \text{ft}$$

So the cheapest tank is 1.65 ft × 3.30 ft × 2.20 ft

Note: This example had a regional constraint because the shape is specified as a rectangle. Also note that when a, b, c are not specified a definite relation other than the product is 12 ft³ yields another solution of a cube 2.29 ft on a side. The functional constraints such as plate stresses due to monitoring of the tank and discontinuity stresses for the vertical and horizontal plates intersecting are missing. Further note, nothing is stated about what the tank holds.

EXAMPLE 3.2. Find the value of R for maximum power Fig. 3.1 being transmitted to R.

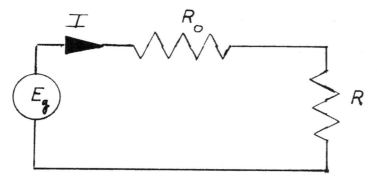

Figure 3.1 Model for power transmission.

R_o–internal resistance of the generator plus the leads.

Criterion function

$$P_R = I^2 R$$

Functional constraint

$$I = \frac{E_g}{R_o + R}$$

Substituting for I

$$P = \frac{E_g^2 R}{(R_o + R)^2}$$

$$\frac{dP}{dR} = \left[\frac{(R_o + R)^2 - R[2(R_o + R)]}{(R_o + R)^4} \right] E_g^2 = 0$$

$$\frac{dP}{dR} = 0 = E_g^2 \frac{(R_o + R)}{(R_o + R)^4} [(R_o + R) - 2R]$$

$$\text{or } R = R_o$$

which says 50% efficiency but maximum delivery of power. This is not practical from a cost basis, however; the communication engineer must deliver maximum power.

There is also another method for using functional constraints with a criterion function.

IV. LAGRANGIAN MULTIPLIERS

A criterion function is known

$$C = c(x_1 \cdots x_n) \tag{3.1}$$

and functional constraints are

$$\begin{aligned} F_1 &= f_1(x_1 \cdots x_n) = 0 \\ \text{to } F_m &= f_m(x_1 \cdots x_n) = 0 \end{aligned} \tag{3.2}$$

If C is to be optimized the total differential [3.22] is developed

$$dC = \frac{\partial C}{\partial x_i} dx_i \cdots + \frac{\partial C}{\partial x_n} dx_n \tag{3.3}$$

or

$$dC = \sum_{i=1}^{i=n} \frac{\partial C}{\partial x_i} dx_i = 0 \tag{3.4}$$

Also Eqs. (3.2) can be differentiated and multiplied by λ_i, the Langrangian multiplier.

$$\lambda_1 dF_1 = \sum_{i=1}^{i=n} \lambda_1 \frac{\partial F_1}{\partial x_i} dx_i = 0$$

to

$$\lambda_m dF_m = \sum_{i=1}^{i=n} \lambda_m \frac{\partial F_m}{\partial x_i} dx_i = 0 \tag{3.5}$$

add Eqs. (3.4) and (3.5)

$$dC + \lambda_1 dF_1 + \cdots \lambda_m dF_m = 0 \tag{3.6}$$

or

$$\sum_{i=1}^{i=n} \left(\frac{\partial C}{\partial x_i} + \lambda_1 \frac{\partial F_1}{\partial x_i} + \cdots + \lambda_m \frac{\partial f_m}{\partial x_i} \right) dx_i = 0 \tag{3.7}$$

Since dx_is are independent and not zero the bracket portions are zero or

$$\frac{\partial C}{\partial x_i} + \lambda_i \frac{\partial F_1}{\partial x_i} + \cdots + \lambda_m \frac{\partial F_m}{\partial x_i} = 0 \quad i = 1, 2, 3, \ldots, n \tag{3.8}$$

One of the advantages of Lagrange's method is that it does not require one to make a choice of independent variables. This is sometimes important in a complex problem. The Lagrange multipliers are often used to verify Kuhn–Tucker necessary conditions [3.22,3.25] for more complex computer optimization. The conditions are incorporated into the computer program and most users are unaware of them. The user's main concern is how to formulate the criterion function, function constraints, and the regional constraints that are bounded to obtain a reasonable solution. Also, are all functions continuous and in a form a computer will work with?

A simple Lagrangian multiplier example is presented.

EXAMPLE 3.3.[3.22]. Find the dimensions of the box of largest volume which can be filled inside an ellipsoid;

$$\frac{x^2}{a^2} + \frac{y^2}{b^2} + \frac{z^2}{c^2} = 1 \tag{3.9}$$

The sides of the box are to be

$2x, 2y, 2z$ (for the sake of symmetry)

so the criterion function C is

$$V = 8xyz \tag{3.10}$$

and the functional constraint is

$$F_1 = \frac{x^2}{a^2} + \frac{y^2}{b^2} + \frac{z^2}{c^2} - 1 = 0 \tag{3.11}$$

Using Eq. (3.8) there is only one λ yielding

$$\frac{\partial c}{\partial x} + \lambda_1 \frac{\partial F_1}{\partial x} = 8yz + \lambda_1 \frac{2x}{a^2} = 0 \tag{3.12}$$

$$\frac{\partial c}{\partial y} + \lambda_1 \frac{\partial F_1}{\partial y} = 8x + \lambda_1 \frac{2y}{b^2} = 0 \tag{3.13}$$

$$\frac{\partial c}{\partial z} + \lambda_1 \frac{\partial F_1}{\partial z} = 8xy + \lambda_1 \frac{2z}{c^2} = 0 \tag{3.14}$$

Now divide by 2 and multiply in order the Eqs. (3.12)–(3.14) by x, y, z.

$$4xyz + \lambda_1 \frac{x^2}{a^2} = 0 \tag{3.15}$$

$$4xyz + \lambda_1 \frac{y^2}{b^2} = 0 \tag{3.16}$$

$$4xyz + \lambda_1 \frac{z^2}{c^2} = 0 \tag{3.17}$$

Adding Eqs. (3.15)–(3.17) and noting the last term is equal to 1

$$12xyz + \lambda_1 \left(\frac{x^2}{a^2} + \frac{y^2}{b^2} + \frac{z^2}{c^2} \right) = 0 \tag{3.18}$$

Substituting λ_1 separately into Eqs. (3.12)–(3.14)

$$2x = \pm \frac{2a}{\sqrt{3}} \qquad 2y = \pm \frac{2b}{\sqrt{3}} \qquad 2z = \pm \frac{2c}{\sqrt{3}} \tag{3.19}$$

The regional constraint requires that only positive values are used.

V. OPTIMIZATION WITH NUMERICAL METHODS

Many times a problem [3.7,3.23] becomes so complex computer numerical iterations are required for a solution.

EXAMPLE 3.4. A hot-water pipe line [3.7,3.23] is to be designed to carry a large quantity of hot water from the heater to the point of use. The cost in dollars per length consists of four items. In this case, only positive values are desired.

(a) Cost of pumping the water from pipe pressure losses

$$C_p = K_p \frac{1}{D^5} = \frac{100}{D^5} \tag{3.20}$$

(b) Cost of heat lost from pipe from the heat transfer through the wall

$$C_h = \frac{K_a}{\ln[(D + 2x)/D]} = \frac{1}{\ln[(D + 2x)/D]} \tag{3.21}$$

(c) Cost of pipe

$$C_{\text{pipe}} = K_3 D = 0.50\,D \tag{3.22}$$

(d) Cost of insulation

$$C_i K_4 x = 1.0\,x \tag{3.23}$$

When the costs are summed

$$C = C_p + C_h + C_{\text{pipe}} + C_i \tag{3.24}$$

The optimal solution [3.7] yields

$D = 1.86$ inches and $x = 1.37$ inches with $C = 4.56$ dollars per length

VI. LINEAR OPTIMIZATION WITH FUNCTIONAL CONSTRAINTS

A criterion function is

$$C = C(x_1 \cdots x_n) \qquad (3.25)$$

with linear functional constraints

$$R_1 \leq F_1(x_1 \cdots x_n) \leq R_1'$$
$$\vdots \qquad\qquad \vdots \qquad (3.26)$$
$$R_m \leq F_m(x_1 \cdots x_n) \leq R_m'$$

When Eqs. (3.25) and (3.26) are linear, it means sums of single power variables. This condition is called linear programming. It should also be realized that a regional constraint is

$$x_1, x_2, \ldots, x_n \geq 0 \qquad (3.27)$$

It has been found that the optimum solution is found at the corners defined by the regional constraints. The constraints in two variables can be easily handled by plotting on graph paper, however, for three or more variables a simplex method is used. Most of the linear programming problems involve mixing, production scheduling, and transportation. These problems tend to be industrial process, chemical, or civil engineering in nature. A few comments are in order about the simplex method.

A. Simplex method [3.14]

The simplex method makes use of the fact an optimum solution is obtained in the corners of the region defined by the regional constraints. The following process is followed for two or three variables.

1. Select a corner of the region as a starting point. The farthest the criterion function is translated from the origin, will yield a minimum or maximum.
2. Choose an edge through this corner such that C increases in value along the edge.
3. Proceed along the edge of the next corner.
4. Repeat steps (2) and (3) until an optimum solution is reached.
5. If a function constraint is parallel to the criterion function any point on the function constraint line yields the same value for the criterion function.

The five step outline is the same as

$$C = c_1x_1 + c_2x_2 + c_3x_3 + \cdots + c_nx_n \tag{3.28}$$

and the functional constraints which bound the solution of the criterion function are

$$a_1x_1 + a_2x_2 + \cdots + a_nx_n = b_1$$
$$a_{n1}x_1 + a_{n2}x_2 + \cdots + a_{nn}x_n = b_n \tag{3.29}$$

with regional constraints for positive values

$$x_n \geq 0 \quad n = 1, 2, \ldots n \tag{3.30}$$

There are other conditions which arise in an actual programming of the above equations.

A simple example of the simplex method in two dimensions follows.

EXAMPLE 3.5 [3.14]. Optimize a two variable criterion function

$$F = x - 2y + 4 \tag{3.31}$$

with the functional constraints

$$x + y \leq 4$$
$$x + 2y \geq -2$$
$$x - y \geq -2 \tag{3.32}$$
$$x \leq 3$$

Plot the functional constraints in Fig. 3.2 and follow the five step simplex method steps outlined. Start at point B where $F = -1$ (step 1) and move along the two edges (step 2) to corner A where $F = 2$ and C where $F = 5$. Now with steps 2 and 3 travel along the edges from A to D and C to D where at D, $F = 12$. The five steps can be used in another fashion.

(a) Plot the criterion function through the origin where $F = 4$ then
(b) Take perpendicular distances d_1 and d_2 where the largest translation of the criterion function is a maximum or minimum. In this case $F = -1$ at B which happens to be a minimum and at D, $F = 12$ the desired maximum.
(c) Also note statement 5 where a functional constraint is parallel to the function constraint. As seen from Fig. 3.2 along any or the parametric lines for F the value of F is constant.

As can be seen corners A and C need not be evaluated and in fact since $d_2 > d_1$ corner D is the only corner to be evaluated but one does not know where the regional minimum is located.

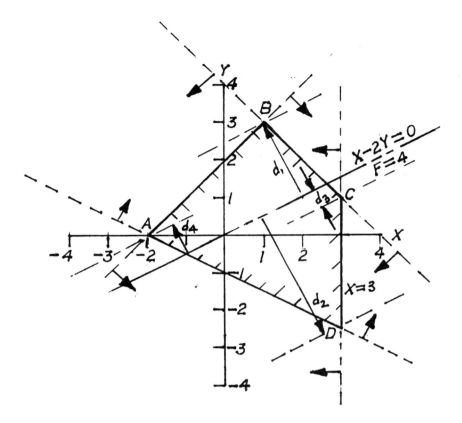

Figure 3.2 Applicable region for example 3.5.

This graphical approach can be extended to three variables but the visual problems with three dimensions are sometimes difficult and a linear programming computer solution saves time. A computer solution is definitely required for more than three variables.

VII. NONLINEAR PROGRAMMING

Linear programming is a function of sum of variables which have single powers like x, y, z. In nonlinear programming the variables have powers and can be products, such as x^3, y^2, $\sqrt{z}$ or xy, y^2z, $x^3\sqrt{z}$. Many engineering problem definitions fall into this category. The following example is also reworked later using geometric programming in Example 3.11 [3.12].

EXAMPLE 3.6. Find the minimum surface area of an open tank with volume no less than 1 unit. The radius is, r, and the height is h.

(a)   The criterion function is the area of the tank

$$g_o(x) = \pi r^2 + 2\pi rh \tag{3.33}$$

(b)   The functional constraint is that of the volume

$$g_1(x) = \pi r^2 h \geq 1 \tag{3.34}$$

The constraint is rewritten

$$g_1(x) \text{ is } \pi r^2 h - 1 \geq 0 \tag{3.35}$$

This is done to comply with the format the computer accepts as input. This must be studied carefully. The answer for r and h are shown in Example 3.11. Always try a problem with known answers to check a new or questionable computer routine. Another example sets up the equations for a nonlinear optimization problem.

(c)   There is a regional constraint in here as the shape of the tank is round

EXAMPLE 3.7. A steel spherical tank holds 250 gallons and is fabricated in two hemispheres and welded to two flanges which are bolted together. The steel for the two halves (neglecting the flanges) is 0.50 dollars per cubic inch and the weld cost is 1.50 dollars/t around the two flanges. The allowable stress is 15 Kpsi for a thin wall analysis.

(a)   Criterion function

$$\text{Cost} = \text{material cost} + \text{welding cost/2 flanges}$$

$$= 0.50(4\pi R^2 t) + \frac{1.50}{t}(2\pi R) \tag{3.36}$$

$$g_o = 2\pi R^2 t + \frac{3\pi R}{t}$$

(b)   Functional constraint 1

$$\text{Volume is } \frac{4\pi R^3}{3} \geq 250 \text{ gallon} \left(\frac{231 \text{ in}^3}{\text{gallon}}\right)$$

$$\tag{3.37}$$

$$g_1 = \frac{4\pi R^3}{3} \frac{1}{57,750 \text{ in}^3} \geq 1$$

(c)   Functional constraint 2

$$\sigma = \frac{PR}{t} \leq 15{,}000 \text{ psi} \tag{3.38}$$

(d)   Regional constraint

all variables > 0

This has been solved in Example 3.12 by geometric programming

cost $= \$1807.12$ $R = 23.978''$ $t = 0.25011''$ $P = 312.9$ psig

EXAMPLE 3.8 [3.1].   A spring optimization derivation [3.1] is summarized as the author developed it and then one of the functional constraints is modified to make the derivation for a fatigue type loading. Figure 3.3 and notation is that of the author. The equations are

The spring weight which is the criterion function is

$$W = \frac{\pi^2 \phi \delta G}{32 P_{\max}} \left(\frac{d^6}{D^2} \right) + \frac{\pi^2 \phi Q}{4} (Dd^2) \tag{3.39}$$

The yielding constraint is

$$\frac{16 P_{\max} D^{0.75}}{\pi \tau_y d^{2.75}} \leq 1 \tag{3.40}$$

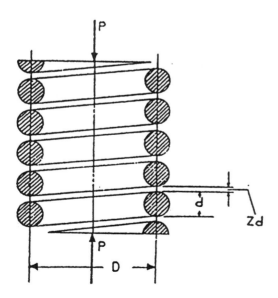

Figure 3.3   Spring cross section.

Harmonic surging constraint

$$\frac{\pi G\delta(13f_d)}{16P_{\max}}\left(\frac{d^3}{D}\right)\sqrt{\frac{32\phi}{Gg}} \leq 1 \tag{3.41}$$

Spring bulking constraint

$$\frac{1.4G\delta^2(1+\frac{v}{2})}{5(1+v)P_{\max}}\left(\frac{d^5}{D^5}\right) \leq 1 \tag{3.42}$$

The author's [3.1] derivation is a lengthy one and a challenge to duplicate.
Nomenclature for Fig. 3.3 and Eqs. (3.39)–(3.42)

$C=$ spring index $= D/d$
$d=$ wire diameter (in)
$z=$ decimal percentage of d, allowance for clearance between adjacent coils
$D=$ mean coil diameter (in)
$\delta=$ deflection corresponding to load $P_{\max}$ (in)
$G=$ torsional modulus of elasticity
$K=$ Wahl factor
$g=$ acceleration of gravity
$n=$ number of active coils
$Q=$ number of inactive coils (end coils)
$P_{\max}=$ maximum spring load (lbs)
$\phi=$ density of spring material (lbs/in^3)
$\tau_y=$ maximum allowable shear stress (psi)
$v=$ Poissons ratio
$f_n=$ natural frequency of fundamental mode of vibrations, cps
$f_d=$ frequency of actuation, cps

The derivation [3.1] is modified for Eq. (3.40) from that of a yielding constraint to one dealing with fatigue. The approach [3.9] is to use a proposed
Wahl failure line [3.24] documented by Faires.

$$\frac{1}{N} = \frac{\tau_m - \tau_a}{S_{ys}} + \frac{2\tau_a}{S_{no}} \tag{3.43}$$

where

$$\tau_m = \frac{8KP_mD}{\pi d^3} \qquad \tau_a = \frac{8KP_aD}{\pi d^3} \tag{3.44}$$

with [3.1]

$$K \approx \frac{2}{C^{0.25}} \tag{3.45}$$

with

$$2 \leq C \leq 12$$

and

$$P_m = \frac{P_{\max} + P_{\min}}{2} \quad P_a = \frac{P_{\max} - P_{\min}}{2} \tag{3.46}$$

with

$$S_{ys} = \frac{\bar{Q}}{d^x} \quad S_{no} = \frac{\bar{Q}'}{d^{x'}} \tag{3.47}$$

The author [3.1] presents a sample problem which is duplicated here for a fatigue condition. A squared and ground spring is subjected to $P_{\max} = 88.2$ lbs and $P_{\min} = \frac{1}{4}$ (88.2 lbs). These are substituted in Eq. (3.46) yields

$$P_m = \frac{P_{\max} + \frac{1}{4} P_{\max}}{2} = \frac{5}{8} P_{\max} \quad P_a = \frac{P_{\max} - \frac{1}{4} P_{\max}}{2} = \frac{3}{8} P_{\max} \tag{3.48}$$

$$\delta_{\max} = 0.5906 \text{ in}, \quad N = 1.15, \quad Q = 2, \text{ and } f_d = 10 \text{ cps}.$$

Table 3.1 is a sample of values found for a variety of materials. The computer solutions found need to be checked for validity.

1. The materials limits need examination.
   (a) Are the answers in the range of the equations used for S_{ys} and S_{no}? ASTM 313 was not!
   (b) Are the values exceeding the maximum stress values for S_{ys} and S_{no}? Here d is substituted into S_{ys}, and S_{no} to verify this. Also the maximum force is used in the stress equation to check again for maximum stress.
   (c) Check all function constraints for the criterion function (the weight).

EXAMPLE 3.9. A two pound steel disk Fig. 3.4 is supported by a round thin wall tube and requires bending and torsional frequencies of greater than 200 Hz each. The minimum thickness for a spring dimension, t, is greater than 0.0015 in, R is less than 2.1211 in so the round spring can be attached and the spring length L is greater than 0.100 in. Since the tube can have thin wall torsional and bending buckling, it must be checked for both. A minimum weight spring is desired. The equations are developed.

Table 3.1 Optimization of spring weight Example 3.8

Spring material [3.9]	G MPSI	S_{ys} KPSI	S_{no} KPSI	W_s lbs.	d inch	D inch	Material equation limits (3.9) S_{ys}	S_{no}	ϕ lb/in³
K-MONEL	9.3	$\dfrac{63.2}{d^{0.048}}$	$\dfrac{18}{d^{0.2}}$	0.839	0.304	2.53	$d \le 0.625$ $S_{ys} \le 72$	$d \le 0.625$ $S_{no} \le 29$	0.306
ASTM 313 COND B S.S.304	10	$\dfrac{79.9}{d^{0.14}}$	$\dfrac{30}{d^{0.17}}$	0.379	0.192	0.946	$0.01 \le d \le 0.130$ $0.13 \le d \le 0.375$ $S_{ys} = (97/d^{0.41})$	$0.01 \le d \le 0.375$ $S_{no} \le 29$	0.29
SPRING BRASS	4.5	$\dfrac{42}{d^{0.26}}$	$\dfrac{11.5}{d^{0.2}}$	1.029	0.289	1.407	$0.08 \le d \le 0.5$ $S_{ys} \le 68$	$0.09 \le d \le 0.5$ $S_{no} \le 19$	0.306
ASTM228 MUSIC WIRE	12	$\dfrac{95}{d^{0.154}}$	$\dfrac{50}{d^{0.154}}$	0.094	0.127	0.605	$0.03 \le d \le 0.192$ $S_{ys} \le 190$	$0.018 \le d \le 0.18$ $S_{no} \le 92$	0.283

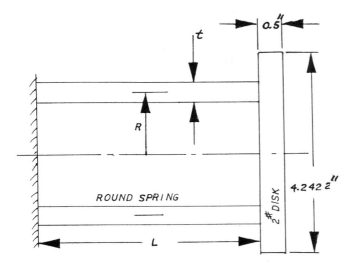

Figure 3.4 Torsional spring cross section.

(a) Criterion function

$$W_{\text{spring}} = 2\pi R t L \gamma \qquad (3.49)$$

There are now four functional constraints

(b) Torsional frequency constraint [3.5]

$$f = \frac{\omega}{2\pi} = \frac{1}{2\pi}\sqrt{\frac{K_\theta}{I_p}}$$

from torsion

$$\frac{TL}{GJ} = \theta$$

$$\frac{T}{\theta} = K_\theta = \frac{GJ}{L}$$

where for thin wall tube the area moment of inertia is

$$J = 2\pi R^3 t$$

The I_p is the mass moment of inertia

$$I_p = \frac{W}{g}\frac{R^2}{2} = 11.6436 \times 10^{-3} \text{ m} \cdot \text{lb} \cdot \text{sec}^2$$

Substituting

$$f = \frac{1}{2\pi}\left[\frac{G2\pi R^3 t}{L}\frac{1}{11.6436 \times 10^{-3}}\right]^{1/2}$$

Now the torsional frequency is

$$f = 3.69714\left[\frac{GR^3 t}{L}\right]^{1/2}$$

The value is to be greater than 200 Hertz so constraint 1, torsional frequency is

$$\frac{200\ \text{Hz}}{3.69714}\left[\frac{L}{GR^3 t}\right] \le 1$$

Constraint 1 torsional frequency

$$54.0958\left[\frac{L}{GR^3 t}\right]^{1/2} \le 1 \qquad\qquad (3.50)$$

(c) Bending frequency constraint [3.5]

$$f_B = \frac{\omega}{2\pi} = \frac{1}{2\pi}\left[\frac{g}{\delta_{ST}}\right]^{1/2}$$

where δ_{ST} is

$$\delta_{ST} = \frac{WL^3}{3EI}$$

Substituting

$$f_B = \frac{1}{2\pi}\left[\frac{3(386.4)}{2\ \text{lb}}\right]^{1/2}\left[\frac{EI}{L^3}\right]^{1/2}$$

$$f_B = 3.83164\left[\frac{EI}{L^3}\right]^{1/2}$$

The I is half of J

$$f_B = 3.83264\left[\frac{E\pi R^3 t}{L^3}\right]^{1/2}$$

f_B is to be greater than 200 hertz

$$\frac{200\ \text{Hz}}{3.83164\sqrt{\pi}}\left[\frac{L^3}{ER^3 t}\right]^{1/2} \le 1$$

Constraint 2 Bending frequency

$$29.449 \left[\frac{L^3}{ER^3 t} \right]^{1/2} \leq 1 \tag{3.51}$$

(d) Torsional buckling [3.21]

$$\tau_{CR} = \frac{0.6E}{\left(\dfrac{2R}{t} \right)^{1.5}}$$

Now $\tau = \sqrt{\dfrac{3}{3}} \sigma$ and σ is to be greater than 10,000 psi

$$\frac{0.6E}{2^{1.5} \left(\dfrac{R}{t} \right)^{1.5}} \frac{3}{\sqrt{3}} \leq 10,000 \text{ psi}$$

$$\frac{0.6}{2^{1.5}} \frac{3}{\sqrt{3}} \frac{1}{10,000} \frac{E}{(R/t)^{1.5}} \leq 1$$

Constraint 3 Torsional buckling

$$36.7423 \times 10^{-6} \frac{E}{\left(\dfrac{R}{t} \right)^{1.5}} \leq 1 \tag{3.52}$$

(e) Bending buckling [3.21]

$$\sigma_{CR} = \frac{0.4E}{\dfrac{2R}{t}}$$

$$\sigma_{CR} = \frac{0.2E}{\dfrac{R}{t}}$$

The left side is 10,000 psi

$$\frac{0.2E}{R/t} \leq 10,000 \text{ psi}$$

Constraint 4 Bending bucking

$$\frac{0.2E}{10,000 R/t} \leq 1 \tag{3.53}$$

There remains three regional constraints, noting R, t, L any of which are zero will make Eq. (3.49) zero weight.

(f)   The radius of the spring must be less than 2.122 in so it can be mounted on the back of the disk.
Constraint 5

$$R \leq 2.1211 \text{ in} \tag{3.54}$$

Also to be fabricated the spring thickness should be more than 0.0015 in
Constraint 6

$$t \geq 0.0015 \text{ in} \tag{3.55}$$

The length must also be developed from some limited dimension (assume 0.100 in or stack up considerations.
Constraint 7

$$L \geq 0.100 \text{ in} \tag{3.56}$$

The Eqs. (3.49)–(3.56) can be placed in a non linear program and values for R, t L found. Use $\gamma = 0.283$ lbs/in^3 and $E = 30 \times 10^6$ psi and compared to Example 3.13.

The spring problem submitted to a nonlinear optimization routine found the following answers:

$$W_s = 0.00024 \text{ lbs}$$
$$R = 0.900 \text{ in}$$
$$t = 0.0015 \text{ in}$$
$$L = 0.100 \text{ in}$$

The answers indicate for a frequency greater than 200 Hz, the length and thickness need a regional constraint from manufacturing considerations.

When the frequency constraints are less than 200 Hz the same optimization routine found:

$$W_s = 0.0102 \text{ lbs}$$
$$R = 0.897 \text{ in}$$
$$t = 0.0015 \text{ in}$$
$$L = 4.281 \text{ in}$$

The geometric programming solution in Example 3.13 using some of the seven constraints in Example 3.9 results in the following values:

$$W_s = 0.00711 \text{ lbs}$$
$$R = 0.900 \text{ in}$$
$$t = 0.0015 \text{ in}$$
$$L = 2.96195 \text{ in}$$

VIII. GEOMETRIC PROGRAMMING

First let's look at some inequalities [3.3,3.8,3.25]

$$(U_1 - U_2)^2 \geq 0$$
$$U_1^2 - 2U_1U_2 + U_2^2 \geq 0 \tag{3.57}$$

add $4U_1U_2$ to both sides

$$U_1^2 + 2U_1U_2 + U_2^2 \geq 4U_1U_2$$

Now take the square root

$$U_1 + U_2 \geq 2U_1^{1/2}U_2^{1/2}$$

Divide by 2 yields

$$\frac{1}{2}U_1 + \frac{1}{2}U_2 \geq U_1^{1/2}U_2^{1/2} \tag{3.58}$$

consider four non-negative numbers

$$\frac{1}{4}U_1 + \frac{1}{4}U_2 + \frac{1}{4}U_3 + \frac{1}{4}U_4 \geq \left(\frac{U_1 + U_2}{2}\right)^{1/2}\left(\frac{U_3 + U_4}{2}\right)^{1/2} \tag{3.59}$$

On the right hand side use Eq. (3.58)

$$U_1 + U_2 \geq 2U_1^{1/2}U_2^{1/2}$$
$$U_3 + U_4 \geq 2U_3^{1/2}U_4^{1/2}$$

So Eq. (3.59) becomes

$$\frac{1}{4}U_1 + \frac{1}{4}U_2 + \frac{1}{4}U_3 + \frac{1}{4}U_4 \geq (U_1^{1/2}U_2^{1/2})^{1/2}(U_3^{1/2}U_4^{1/2})^{1/2}$$
$$\geq U_1^{1/4}U_2^{1/4}U_3^{1/4}U_4^{1/4} \tag{3.60}$$

Zener made the observation if one lets

$$u_i = \delta_i U_i \quad \text{then Eq. (3.60)} \tag{3.61}$$

and

$$U_i = u_i/\delta_i$$

$$u_1 + u_2 + u_3 + u_4 \geq \left(\frac{u_1}{\delta_1}\right)^{\delta_1}\left(\frac{u_2}{\delta_2}\right)^{\delta_2}\left(\frac{u_3}{\delta_3}\right)^{\delta_3}\left(\frac{u_4}{\delta_4}\right)^{\delta_4} \tag{3.62}$$

Now if this is true allow $u_2 = u_3 = u_4$ in Eq. (3.60) then

$$\frac{1}{4}U_1 + \frac{3}{4}U_4 \geq U_1^{1/4}U_4^{3/4} \tag{3.63}$$

or

$$\delta_1 U_1 + \delta_4 U_4 \geq U_1^{\delta_1}U_4^{\delta_4}$$

and using Eq. (3.62)

$$u_1 + u_4 \left(\frac{u_1}{\delta_1}\right)^{\delta_1}\left(\frac{u_4}{\delta_4}\right)^{\delta_4} \tag{3.64}$$

Now the general expression for geometric programming is

$$\delta_1 U_1 + \delta_2 U_2 + \delta_3 U_3 + \delta_4 U_4 \cdots + \delta_n U_n \geq U_1^{\delta_1}U_2^{\delta_2} \cdots U_n^{\delta_n} \tag{3.65}$$

Using Eq. (3.61)

$$u_1 + u_2 + u_3 + \cdots u_n \geq \left(\frac{u_1}{\delta_1}\right)^{\delta_1}\left(\frac{u_2}{\delta_2}\right)^{\delta_2} \cdots \left(\frac{u_n}{\delta_n}\right)^{\delta_n} \tag{3.66}$$

Now in Eq. (3.65) if all Us are equal

$$\delta_1 + \delta_2 + \delta_3 + \cdots \delta_n \geq 1 \quad \text{or}$$
$$\sum \delta_n = 1 \text{ for a minimum} \tag{3.67}$$

So the equations are

$$\sum \delta_i U_i \geq \prod U_i^{\delta_i} \tag{3.68}$$

$$\sum \delta_i = 1 \tag{3.69}$$

or

$$u_i = \delta_i U_i$$
$$\sum u_i \geq \prod \left(\frac{u_i}{\delta_i}\right)^{\delta_i} \tag{3.70}$$

EXAMPLE 3.10. Minimize the following function [3.3,3.8]

$$g = \frac{\beta_1}{x_1 x_2 x_3} + \beta_2 x_2 x_3 + \beta_3 x_1 x_3 + \beta_4 x_1 x_2 \tag{3.71}$$

use the form Eq. (3.70)

$$\sum u_i \geq \prod \left(\frac{u_i}{\delta_i}\right)^{\delta_i}$$

$$\sum \delta_i = 1$$

$$u_1 = \frac{\beta_1}{x_1 x_2 x_3} \qquad u_2 = \beta_2 x_2 x_3$$

$$u_3 = \beta_3 x_1 x_3 \qquad u_4 = \beta_4 x_1 x_2$$

(3.72)

So

$$\frac{\beta_1}{x_1 x_2 x_3} + \beta_2 x_2 x_3 + \beta_3 x_1 x_3 + \beta_4 x_1 x_2 \geq$$

$$\left(\frac{\beta_1}{\delta_1 x_1 x_2 x_3}\right)^{\delta_1} \left(\frac{\beta_2 x_2 x_3}{\delta_2}\right)^{\delta_2} \left(\frac{\beta_3 x_1 x_3}{\delta_3}\right)^{\delta_3} \left(\frac{\beta_4 x_1 x_2}{\delta_4}\right)^{\delta_4}$$

(3.73)

$$\delta_1 + \delta_2 + \delta_3 + \delta_4 = 1 \tag{3.74}$$

Look at the right hand side of Eq. (3.73) and rearrange

$$g \geq \left(\frac{\beta_1}{\delta_1}\right)^{\delta_1} \left(\frac{\beta_2}{\delta_2}\right)^{\delta_2} \left(\frac{\beta_3}{\delta_3}\right)^{\delta_3} \left(\frac{\beta_4}{\delta_4}\right)^{\delta_4} x_1^{-\delta_1+\delta_3+\delta_4} x_2^{-\delta_1+\delta_2+\delta_4} x_3^{-\delta_1+\delta_2+\delta_3} \tag{3.75}$$

A minimum is obtained if

$$x_1^{-\delta_1+\delta_3+\delta_4} x_2^{-\delta_1+\delta_2+\delta_4} x_3^{-\delta_1+\delta_2+\delta_3} = 1 \tag{3.76}$$

or the exponents are zero

$$-\delta_1 + 0 + \delta_3 + \delta_4 = 0$$

$$-\delta_1 + \delta_2 + 0 + \delta_4 = 0$$

$$-\delta_1 + \delta_2 + \delta_3 + 0 = 0$$

$$\delta_1 + \delta_2 + \delta_3 + \delta_4 = 1$$

(3.77)

and in Eq. (3.77)

$$\delta_1 = 2/5 \quad \delta_2 = 1/5 \quad \delta_3 = 1/5 \quad \delta_4 = 1/5$$

Examine the degrees of difficulty

$$DD = T - (N + 1)$$

$$T = \text{Number of terms}$$

$$N = \text{Number of variable}$$

(3.78)

From Eq. (3.71) Example 3.10

    4 terms in the original expression

    3 variables

    $DD =$ zero

This means the equation can be solved as an algebra problem. However, if $DD>$zero this becomes an interation problem for a geometric programming optimization routine. The next example contains constraints and is more difficult.

    EXAMPLE 3.11. Look at an example from [3.12] expanded from the original text which outlines a method to formulate other problems.

    Find the minimum area of an open cylindrical tank Example 3.6 with volume no less than 1 unit. The radius is r and the height is h.

$$g_0(x) = \pi r^2 + 2\pi rh \qquad \text{area of tank}$$
$$g_1(x) = \pi r^2 h \geq 1 \qquad \text{constant} \tag{3.79}$$

Degree difficulty $= T-(N+1)=3-(2+1)=0$
Let $u_1 = \pi r^2$ and $u_2 = 2\pi rh$ Eq. (3.70) substituted into Eq. (3.79)

$$g_0 \geq \left(\frac{u_1}{\delta_1}\right)^{\delta_1} \left(\frac{u_2}{\delta_2}\right)^{\delta_2}$$

or

$$g_0 \geq \left(\frac{\pi r^2}{\delta_1}\right)^{\delta_1} \left(\frac{2\pi rh}{\delta_2}\right)^{\delta_2} \tag{3.80}$$

In the volume constraint divide by $\dfrac{1}{\pi r^2 h}$ so

$$1 \geq \frac{1}{\pi r^2 h} \quad \text{or}$$

$$g_1 = \frac{1}{\pi r^2 h} \leq 1$$

or placing g_1 in a similar form as g_0 with $u_1^1 = \dfrac{1}{\pi r^2 h}$ or $1 \geq g_1$

$$1 \geq \left[\frac{u_1^1}{\delta_1^1}\right]^{\delta_1^1} \mu_1 \tag{3.81}$$

This obtains a dual objective function
$$V(\delta) = g_0 g_1 \quad \text{or}$$

$$V(\delta) = \left(\frac{\pi r^2}{\delta_1}\right)^{\delta_i} \left(\frac{2\pi r h}{\delta_2}\right)^{\delta_2} \left[\left(\frac{1}{\pi r^2 h}\right)\left(\frac{1}{\delta_1^1}\right)\right]^{\delta_1^1} \mu_1^{\mu_1}$$

or

$$V(\delta) = \left(\frac{\pi}{\delta_1}\right)^{\delta_1} \left(\frac{2\pi}{\delta_2}\right)^{\delta_2} \left[\left(\frac{1}{\pi}\right)\left(\frac{1}{\delta_1^1}\right)\right]^{\delta_1^1} \mu_1^{\mu_1} \ r^{(2\delta_1 + \delta_2 - 2\delta_1^1)} h^{(\delta_2 - \delta_1^1)} \tag{3.82}$$

$$V(\delta) \to g_0(x) \text{ for a minimum}$$

Now $V(\delta)$ is a minimum if the powers on r and h are 0 or orthogonality constants for Eq. (3.82)
$$2\delta_1 + \delta_2 - 2\delta_1^1 = 0$$
$$\delta_2 - \delta_1^1 = 0 \tag{3.83}$$

Also Eq. (3.67)
$$\delta_1 + \delta_2 = 1 \tag{3.84}$$

for the constraint, a 2nd normality constraint
$$\delta_1^1 = \mu_1 \tag{3.85}$$

A solution gives
$$\delta_1 = 1/3 \quad \delta_2 = 2/3 \quad \delta_1^1 = 2/3 \quad \mu_1 = 2/3 \tag{3.86}$$

Answers are obtained when $V(\delta)$ is evaluated r^o, h^o are 1
$$V(\delta) = \left(\frac{\pi}{1/3}\right)^{1/3} \left(\frac{2\pi}{2/3}\right)^{2/3} \left[\frac{1}{2\pi/3}\right]^{2/3} (2/3)^{2/3}$$
$$= (3\pi)^{1/3} (3\pi)^{2/3} \left[\frac{3}{2\pi}\frac{2}{3}\right]^{2/3} \tag{3.87}$$
$$V(\delta) = 3\pi \left(\frac{1}{\pi}\right)^{2/3}$$

Now from δ_1
$$\delta_1 = \frac{1}{3} = \frac{U_1}{V(\delta)} = \frac{\pi r^2}{3\pi \left(\frac{1}{\pi}\right)^{2/3}}$$

$$r^2 = \left(\frac{1}{\pi}\right)^{2/3} \tag{3.88}$$

$$r = \left(\frac{1}{\pi}\right)^{1/3}$$

Next from δ_2

$$\delta = 2/3 = \frac{U_2}{V(\delta)} = \frac{2\pi rh}{3\pi\left(\dfrac{1}{\pi}\right)^{2/3}}$$

$$rh = \left(\frac{1}{\pi}\right)^{2/3}$$

Substitute for r Eq. (3.88)

$$h = \left(\frac{1}{\pi}\right)^{1/3} \tag{3.89}$$

Now check if $V(\delta) = g_0$ by substituting for r and h

$$g_0 = \pi r^2 + 2\pi rh$$

$$g_0 = \pi\left[\left(\frac{1}{\pi}\right)^{1/3}\right]^2 + 2\pi\left[\left(\frac{1}{\pi}\right)^{1/3}\right]\left[\left(\frac{1}{\pi}\right)^{1/3}\right] \tag{3.90}$$

$$g_0 = 3\pi\left(\frac{1}{\pi}\right)^{2/3}$$

Yes! they are equal and a minimum has been found. Next check the constraint:

$$\pi r^2 h \geq 1 \tag{3.91}$$

Substituting for r and h

$$\pi\left[\left(\frac{1}{\pi}\right)^{1/3}\right]^2\left[\left(\frac{1}{\pi}\right)^{1/3}\right] \geq 1$$

$$\pi\left(\frac{1}{\pi}\right)^{2/3}\left(\frac{1}{\pi}\right)^{1/3} \geq 1$$

$$\frac{\pi}{\pi} = 1$$

constraint is satisfied.

EXAMPLE 3.12. From Example 3.7
Criterion function

$$\text{Cost} = 2\pi R^2 t + \frac{3\pi R}{t} \tag{3.92}$$

Functional constraints

Volume

$$\frac{4\pi R^3}{3} \geq 57{,}750 \text{ in}^3 \tag{3.93}$$

Stress in a sphere

$$\frac{PR}{2t} \leq 15{,}000 \text{ psi} \qquad (3.94)$$

Regional constraint

Variables

$$\begin{aligned} R &> 0 \\ t &> 0 \\ P &> 0 \end{aligned} \qquad (3.95)$$

In the cost let

$$U_1 = 2\pi R^2 t \qquad U_2 = \frac{3\pi R}{t} \qquad (3.96)$$

Volume

$$\begin{aligned} 1 &\geq \frac{3(57{,}750 \text{ in}^3)}{4\pi R^3} \\ U_1^1 &= \frac{3(57{,}750)}{4\pi R^3} \end{aligned} \qquad (3.97)$$

Stress

$$\frac{PR}{2t(15{,}000)} \leq 1 \qquad (3.98)$$

$$\begin{aligned} U_1'' &= \frac{PR}{2(15{,}000t)} \\ g_0 &= \left(\frac{2\pi R^2 t}{\delta_1}\right)^{\delta_1} \left(\frac{3\pi R}{t\delta_2}\right)^{\delta_2} \end{aligned} \qquad (3.99)$$

$$g_1 = \left[\left(\frac{3(57{,}750)}{\delta_1^1 \, 4\pi R^3}\right)^{\delta_1^1} \right] \mu_1^{\mu_1} \qquad (3.100)$$

$$g_2 \text{ is } \left[\left(\frac{PR}{\delta_1''(30{,}000)t}\right)^{\delta_1''} \right] \mu_2^{\mu_2} \qquad (3.101)$$

Now create the dual objective function

$$V(\delta) = g_0 g_1 g_2 \tag{3.102}$$

$$D.D. = T - (N + 1) = 4 - (3 + 1) = 0$$

Substituting

$$V(\delta) = \left(\frac{2\pi R^2 t}{\delta_1}\right)^{\delta_1} \left(\frac{3\pi R}{\delta_2 t}\right)^{\delta_2} \left[\left(\frac{3(57,750)}{4\pi R^3 \delta_1^1}\right)^{\delta_1^1}\right]^{\mu_1} \left[\left(\frac{PR}{30,000 t \delta_1''}\right)^{\delta_1''}\right]^{\mu_2}$$

Now combining all the variables

$$V(\delta) = \left(\frac{2\pi}{\delta_1}\right)^{\delta_1} \left(\frac{3\pi}{\delta_2}\right)^{\delta_2} \left[\left(\frac{3(57,750)}{4\pi \delta_1^1}\right)^{\delta_1^1}\right]^{\mu_1} \left[\left(\frac{1}{30,000 \delta_1''}\right)^{\delta_1''}\right]^{\mu_2}$$
$$\times P^{\delta_1''} R^{2\delta_1 + \delta_2 - 3\delta_1^1 + \delta_1''} t^{\delta_1 - \delta_2 - \delta_1''}$$

$$\tag{3.103}$$

Now first orthogonalities

$$\delta_1 + \delta_2 = 1 \tag{3.104}$$

$$\delta_1^1 = \mu_1 \tag{3.105}$$

$$\delta_1'' = \mu_2 \tag{3.106}$$

From powers on P, R, t

$$\delta_1'' = 0 \tag{3.107}$$

$$2\delta_1 + \delta_2 - 3\delta_1^1 + \delta_1'' = 0 \tag{3.108}$$

$$\delta_1 - \delta_2 - \delta_1'' = 0 \tag{3.109}$$

Now from inspection from Eq. (3.107) and Eq. (3.106)

$$\delta_1'' = \mu_2 = 0$$

Substituting Eq. (3.107) into Eq. (3.109) and solving Eq. (3.104) and Eq. (3.109)

$$\delta_1 + \delta_2 = 1$$
$$\delta_1 - \delta_2 + 0 = 0$$

Adding

$$2\delta_1 = 1$$
$$\delta_1 = \frac{1}{2}$$
$$\delta_1 = \delta_2 = \frac{1}{2}$$

(3.110)

Substitute Eq. (3.110) and Eq. (3.107) into Eq. (3.108)

$$2\left(\frac{1}{2}\right) + \left(\frac{1}{2}\right) - 3\delta_1^1 + (0) = 0$$
$$\frac{3}{2} - 3\delta_1^1 = 0$$
$$-3\delta_1^1 = -\frac{3}{2}$$
$$\delta_1^1 = 1/2 = \mu_1$$

(3.111)

Now substitute Eq. (3.111) and Eq. (3.110) into the $V(\delta)$ Eq. 3.103

$$V(\delta) = \left(\frac{2\pi}{1/2}\right)^{1/2} \left(\frac{3\pi}{1/2}\right)^{1/2} \left[\left(\frac{3(57,750)}{4\pi/2}\right)^{1/2}\right](1/2)^{1/2}$$
$$\times \left[\left(\frac{1}{15,000(0)}\right)^0\right]0^0 p^0 R^0 t^0$$

(3.112)

The terms to the zero power are equal to 1

$$a^0 = 1 \quad \text{if} \quad a \neq 0$$

The last constraint due to stress is not binding and can be dropped out of the problem formulation as it really doesn't affect the cost. Thus, evaluating the $V(\delta)$ minimum cost without the last constraint

$$V(\delta) = \left[(4\pi)(6\pi)\frac{(6[57,750])}{4\pi}\right]^{1/2}\left(\frac{1}{2}\right)^{1/2}$$
$$= [36\pi[57,750]]^{1/2}(1/2)^{1/2}$$
$$= \$1807.12$$

(3.113)

to evaluate variables R and t.

Now from δ_1

$$\delta_1 = \frac{1}{2} = \frac{U_1}{V(\delta)} = \frac{2\pi R^2 t}{1807.12} \tag{3.114}$$
$$R^2 t = 143.806$$

From δ_2

$$\delta_2 = \frac{1}{2} = \frac{U_2}{V(\delta)} = \frac{3\pi R}{t}\frac{1}{1807.12} \tag{3.115}$$
$$\frac{R}{t} = 95.8707$$

From δ_1^1

$$\frac{\delta_1^1}{\mu_1} = U_1^1 = \frac{3(57,750)}{4\pi R^3} = 1$$
$$R^3 = \frac{3(57,750 \text{ in}^3)}{4\pi} \tag{3.116}$$
$$R = 23.9785''$$

Now looking at the stress constraint even when it's not binding.

$$\frac{PR}{2t} \leq 15,000$$

From Eq. (3.115)

$$\frac{P}{2}(95.8707) \leq 15,000 \tag{3.117}$$
$$P \leq 312.921 \text{ psig}$$

This is highest pressure which also allows minimum cost.
Now substitute Eq. (3.116) into Eq. (3.115)

$$\frac{R}{t} = \frac{23.9785}{t} = 95.8707 \tag{3.118}$$
$$t = 0.2501''$$

So

$$\text{Cost} = \$1807.12$$
$$P_{\text{max}} = 312.9 \text{ psig}$$
$$t = 0.2501''$$
$$R = 23.98''$$

EXAMPLE 3.13. Example 3.9 is examined for a geometric programming solution. First collect Eqs. (3.49)–(3.56)

(a) Criterion function

$$W_{\text{spring}} = 2\pi RtL\,\gamma \qquad (3.119)$$

(b) Functional constraint 1 for torsional frequency

$$54.0958\left[\frac{L}{GR^3t}\right]^{1/2} \le 1 \qquad (3.120)$$

(c) Functional constraint 2 for bending frequency

$$29.449\left[\frac{L^3}{ER^3t}\right]^{1/2} \le 1 \qquad (3.121)$$

(d) Functional constraint 3 for torsional buckling

$$36.7423 \times 10^{-6}\,\frac{E}{(R/t)^{1.5}} \le 1 \qquad (3.122)$$

(e) Functional constraint 4 for bending buckling

$$2 \times 10^{-5}\,\frac{E}{\left(\dfrac{R}{t}\right)} \le 1 \qquad (3.123)$$

(f) Regional constraint 5 for spring mean radius

$$R \le 2.122 \text{ in} \qquad (3.124)$$

(g) Regional constraint 6 for spring thickness

$$t \ge 0.0015 \text{ in} \qquad (3.125)$$

(h) Regional constraint for spring length

$$L \ge 0.100 \text{ in} \qquad (3.126)$$

Before starting examine the degrees of difficulty Eq. (3.78) where for selected E and γ

Terms $= 7$ Variables (3)R, t, L

$DD = 7 - (3 + 1) = 3$

The ideal DD is zero and if three of the constraints could be left out a hand solution it could be solved. However, a check of all constraints at the end for values of R, t, and L must hold. Constraints 5 and 6 give an indication of where the answer for R and t should be; so ignore these but start solving for values using them. Then constraints 3 and 4 are similar for R and t and some selected values indicate the bending constraint 4 is to be selected. Constraint 4 is needed so the spring maintains stability. Constraint 3 will be ignored but definitely checked at the end.

The modulus, E, and the density, γ, are related [3.13] from vibration E/γ the specific stiffness

$$\frac{E}{\gamma} = 105 \times 10^6 \text{ in} \tag{3.127}$$

for most common structural members. This could introduce another variable to solve for, but, what does one do when the solved value for E does not exist in any known material. The best method is to introduce the known discrete values of E and γ for common materials.

The relationship for E and G [3.21] is

$$G = \frac{E}{2(1 - v)} = \frac{E}{2.6} \tag{3.128}$$

Now substitute $E = 30 \times 10^6$ psi and $\gamma = 0.283$ lb/in^3 into Eqs. (3.119)–(3.123) and (3.128). Note: If a known spring material is used more precise numbers are available. The equation for solution are

Spring weight, g_0

$$W_{\text{spring}} = 1.77814 \, Rtl = A \, RtL \tag{3.129}$$

Constraint 1 f_T

$$15.9254 \times 10^{-3} \left[\frac{L}{R^3 t} \right]^{1/2} \leq 1$$

$$B \left[\frac{L}{R^3 t} \right]^{1/2} \leq 1 \tag{3.130}$$

Constraint 2 f_B

$$5.37663 \times 10^{-3} \left[\frac{L^3}{R^3 t} \right]^{1/2} \leq 1$$

$$C \left[\frac{L^3}{R^3 t} \right]^{1/2} \leq 1 \tag{3.131}$$

Constraint 4 σ_{CRB}

$$600 \left(\frac{t}{R} \right) \leq 1$$

$$D \left(\frac{t}{R} \right) \leq 1 \tag{3.132}$$

Following Example 3.12 in Eq. (3.129)

$$U_1 = ARtL$$

then

$$g_0 = \left(\frac{U_1}{\delta_1}\right)^{\delta_1} \tag{3.133}$$

In Eq. (3.130)

$$U_1^1 = B\left[\frac{L}{R^3 t}\right]^{1/2}$$
$$g_1 = \left(\frac{U_1^1}{\delta^1}\right)^{\delta_1^1} \tag{3.134}$$

In Eq. (3.131)

$$U_1'' = C\left[\frac{L^3}{R^3 t}\right]^{1/2}$$

giving

$$g_2 = \left(\frac{U_1'''}{\delta_1''}\right)^{\delta_1''} \tag{3.135}$$

and lastly in Eq. (3.132)

$$U_1''' = D\left(\frac{t}{R}\right)$$
$$g_0 = \left(\frac{U_1'''}{\delta_1'''}\right)^{\delta_1'''} \tag{3.136}$$

The dual objective function is developed

$$V(\delta) = g_0 g_1 g_2 g_3$$

Substituting Eqs. (3.133)–(3.136) and rearranging

$$V(\delta) = \left[\frac{A}{\delta_1}\right]^{\delta_1}\left[\left(\frac{B}{\delta_1^1}\right)^{\delta_1^1}\right]\mu_1^{\mu_1}\left[\left(\frac{C}{\delta_1''}\right)^{\delta_1''}\right]\mu_2^{\mu_2}\left[\left(\frac{D}{\delta_1'''}\right)^{\delta_1'''}\right]\mu_3^{\mu_3}$$
$$R^{\delta_1 - \frac{3}{2}\delta_1' - \frac{3}{2}\delta_1'' - 1\delta_1'''}\, t^{\delta_1 - \frac{1}{2}\delta_1' - \frac{1}{2}\delta_1'' + 1\delta_1'''}\, L^{\delta_1 + \frac{\delta^1}{2} + \frac{3}{2}\delta_1''} \tag{3.137}$$

Now from orthogonality Eq. (3.67) and Eq. (3.85)

$$\sum \delta_1 = 1 \tag{3.138}$$

$$\delta_1 = 1 \text{ and } \delta_1^1 = \mu_1 \tag{3.139}$$

$$\delta_1^1 = \mu_2 \tag{3.140}$$

$$\delta_1''' = \mu_3 \tag{3.141}$$

The powers on R, t, and L are zero for a minimum

$$\delta_1 - 3\delta_1'/2 - 3\delta_1''/2 - \delta_1''' = 0 \tag{3.142}$$

$$\delta_1 - \delta_1'/2 - \delta_1''/2 + \delta_1''' = 0 \tag{3.143}$$

$$\delta_1 + \delta_1'/2 + 3\delta_1''/2 + 0 = 0 \tag{3.144}$$

Solving using Eq. (3.138) and Eqs. (3.142)–(3.144)

$$\delta_1 = 1 \quad \delta_1^1 = 5/2 \quad \delta_1'' = -3/2 \quad \delta_1''' = -1/2 \tag{3.145}$$

Now to substitute δ values Eq. (3.145) into Eq. (3.137) yields

$$V(\delta) = 0.00589 \text{ lbs}$$

This will be the weight of the spring g_0 Eq. (3.129) and (3.133) if optimum values for R, t and L can be found.

The equations to size the dimensions are developed from Eqs. (3.129)–(3.132).

$$\delta_1 = \frac{U_1}{V(\delta)} = 1 = \frac{1.77814 R t L}{0.00589} \tag{3.146}$$

$$\frac{\delta_1'}{\mu_1} = U_1' = 1 = 15.9253 \times 10^{-3} \left[\frac{L}{R^3 t}\right]^{1/2} \tag{3.147}$$

$$\frac{\delta_1''}{\mu_2} = U_1'' = 1 = 5.37663 \times 10^{-3} \left[\frac{L^3}{R^3 t}\right]^{1/2} \tag{3.148}$$

$$\frac{\delta_1''}{\mu_3} = U_1''' = 1 = 600(t/R) \tag{3.149}$$

Constraint 3 is not used but will be checked with Eqs. (3.119)–(3.126)

Constraint 5 $R \leq 2.1211$ in
Constraint 6 $t \geq 0.0015$ in
Constraint 7 $L \geq 0.100$ in

Equate Eq. (3.148) and Eq. (3.147) solving for L yielding

$$L = 2.96195'' \tag{3.150}$$

into Eq. (3.146) substitute Eq. (3.149) for R and Eq. (3.150)

$$t = 0.001366'' \tag{3.151}$$

But constraint 6 states $t \geq 0.0015$ in. from Eq. (3.149), $t = 0.0015''$

$$R = 0.900'' \tag{3.152}$$

Into Eqs. (3.119) substitute Eq. (3.150), (3.152) and t = 0.0015″

$$w_s = g_0 = 0.00711 \text{ lbs} \tag{3.153}$$

$$V(\delta) = 0.00589 \text{ lbs} \tag{3.154}$$

The other constraints Eq. (3.120)–(3.126) must be checked for solutions for R, t, and L.

Constraint 1

$0.828 \leq 1$

Constraint 2

$0.828 \leq 1$

Constraint 3

$0.075 \leq 1$

Constraint 4

$1 \leq 1$

Constraint 5

0.9 in ≤ 2.1211 in

Constraint 6

0.0015 in ≥ 0.0015 in

Constraint 7 for L

$2.96195 \geq 0.100$ in

Note: The equality constraints are the binding equations of the solution. Also, in previous examples $V(\delta)$ and g_0 (the w_s) always equated if all constraints were useable. Here only three of the seven are used hence $V(\delta)$ didn't get the proper feed back from all seven constraints. From the check of constraints, constraints 4, 6, 7 are more important.

REFERENCES

3.1.   Agrawal GK. Optimal Design of Helical Springs for Minimum Weight by Geometric Programming, ASME 78-WA/DE-1, 1978.

3.2.   Aoki M. Optimization Techniques, MacMillian, 1971.

3.3.   Converse AO. Optimization, New York: Holt, Rinehart and Winston, 1970.

3.4.   Bain LJ. Statistical Analysis of Reliability and Life-Testing Models (Theory and Methods), Marcel Dekker, 1978.

3.5.   Den Hartog JP. Mechanical Vibrations, 4th ed, New York: McGraw-Hill Book Co. 1956.

3.6.   DiRoccaferrera GMF. Introduction to Linear Programming, South-Western, 1967.

3.7.   Dixon JR. Design Engineering, New York: McGraw-Hill Book Co, 1966.

3.8.   Duffin RJ, Peterson EL, Zener C. Geometric Programming, New York: Wiley, 1967. Also a later 2nd ed.

3.9.   Faires VM. Design of Machine Elements, 4th ed, New York: MacMillian Company, 1965.

3.10.   Fox RL. Optimization Methods for Engineering Design, Addison-Wesley, 1971.

3.11.   Furman TT. Approximate Methods in Engineering Design, Academic Press, 1981.

3.12.   Gottfried BS, Weisman J. Introduction to Optimization Theory, Englewood Cliffs, NJ: Prentice-Hall, 1973.

3.13.   Griffel W. Handbook of Formulas for Stress and Strain, New York: Frederick Ungar Publishing Co, 1966.

3.14.   Lipschutz S. Finite Mathematics, New York: McGraw-Hill, 1966.

3.15.   Mann NR, Schafer RE, Sing Purwalla ND. Methods for Statistical Analysis of Reliability and Life Data, New York: John Wiley and Sons, 1974.

3.16.   Mechanical Engineering News, Vol. 7, No. 2, May 1970.

3.17.   Peters MS. Plant Design and Economics for Chemical Engineers, New York: McGraw-Hill, 1958.

3.18.   Reeser C. Making Decisions Scientifically, 29 May 1972, Machine Design, Cleveland: Penton Co, 1972.

3.19.   Rubenstein R. Simulation and the Monte Carlo Method, New York: Wiley and Sons, 1981.

3.20.   SAS/IML Software Changes and Enhancements Through Release 6.11, SAS Institute Inc, Cary NC, 1995.

3.21.   Shanley FR. Strength of Materials, New York: McGraw-Hill Book Co, 1957.

3.22. Taylor A. Advanced Calculus, Ginn and Co, 1955.
3.23. Vidosic JP. (1969) Elements of Design Engineering, New York: The Ronald Press Co, 1969.
3.24. Wahl AM. Variable Stresses in Springs, January–April 1938 Machine Design, Penton Co., Cleveland.
3.25. Wilde DJ, Beightler CS, Foundations of Optimization, Englewood Cliffs, NJ: Prentice-Hall, 1967. Also a later 2nd ed.
3.26. Yasak T. A method of minimum weight design with requirements imposed on stresses and natural frequencies, Report 452, Institute of Space and Aeronautical Science, University of Tokyo, 1970.

PROBLEMS

PROBLEM 3.1

Find the dimensions of the largest area rectangle that can be inscribed in a circle with a radius of 10 ft.

PROBLEM 3.2

Storage containers are to be produced having a volume of 100 cubic feet each. They are to have a square base and an open top. What dimensions should the container have in order to minimize the amount of material required (i.e. minimize the cost)?

PROBLEM 3.3

A manufacturer produces brass bolts and mild steel bolts at an average cost of 40¢ and 20¢, respectively. If the brass bolts are sold for X cents and the mild steel bolts are sold for Y cents, the market per quarter is $4,000,000/XY$ brass bolts and $8,000,000/XY$ mild steel bolts. Find the selling prices for maximum profit.

PROBLEM 3.4

A thin wall cantilever tube with a thickness t greater than 0.002 in and 60 in long is loaded at the tip with a 500 lbs load offset 6 in. perpendicular from the center of the tube. The material is 6061-T6 aluminum with an allowable tension of 31,900 psi. Include the torsional buckling and bending perpendicular buckling constraints. Solve for the minimum weight of the tube using nonlinear or geometric programming.

PROBLEM 3.5

An open conveyor bucket with dimensions on the right triangular cross section of $2h$ width by h height and L length has a capacity of one cubic foot. The cost for material is five dollars per square foot and welding is ten dollars per linear foot. Find the dimensions to produce a minimum cost conveyor bucket.

PROBLEM 3.6

A toy manufacturer makes two types of plastic boats. The manufacturing data are as follows:

Process	Production time req.		Available time
	X	Y	
Molding	10	5 min	80
Sanding and painting	6	6	66
Assembling	5	6	90
Profit per unit	$1.20	$1.00	

Find the production rates of the two models which will maximize profit.

PROBLEM 3.7

A container manufacturer produces two types of boxes. The production requirements are as follows:

Machine	Mfg. time required, min per unit		Available capacity per time period, min.
	Box A	Box B	
1	4.0	2.0	2,000
2	3.0	5.0	3,000
Profit per unit	20¢	10¢	

Find the production rates for maximum profit.

PROBLEM 3.8

A casting company wishes to know the production of products (P_1, P_2, P_3, P_4, P_5, P_6) in Table Problem 3.8 which will give a maximum profit. The operations are shown below.

Table Problem 3.8

Operation	Available Time/Wk	Operating Cost	Product Time/Unit(min.)					
			P_1	P_2	P_3	P_4	P_5	P_6
M_1 casting	2200 min	$0.15/min	8	3	4	5	6	2
M_2 deburring	2400 min	$0.08/min	4	4	6	2	3	2
M_3 drilling	2400 min	$0.17/min	0	1	2	8	4	3
M_4 tapping	400 min	$0.09/min	4	6	2	0	0	0
M_5 drilling	400 min	$0.012/min	1	5	2	0	0	0
	Material cost		0.80	0.65	0.30	0.40	0.45	1.00
	Selling price/unit		7.00	5.50	4.50	5.50	4.30	4.00

Solve for P_1–P_6 to maximize profit

PROBLEM 3.9

A steam plant [3.16] has two boilers which it normally operates. Both are equipped to burn either coal, oil, or gas according to the following efficiencies:

	Coal	Oil	Gas
Boiler 1	0.80	0.82	0.84
Boiler 2	0.60	0.65	0.76

Total steam supply required of the two boilers is 150 heat units/day. Maximum output Boiler 1 is 100 units/day and of Boiler 2 is 90 units/day. A contract requires 120 units of gas/day to be purchased; maximum limit on coal must be 150 heat units/day; and oil must be 20 heat units/day.

	Coal	Oil	Gas
Fuel costs in cents/heat unit	25	27	29

Verify the solution

X_1 coal in Boiler 1
X_2 coal in Boiler 2
X_3 oil in Boiler 1
X_4 oil in Boiler 2
X_5 gas in Boiler 1
X_6 gas in Boiler 2

Cost/Day:

$$Y = 25X_1 + 25X_2 + 27X_3 + 27X_4 + 29X_5 + 29X_6$$

1. Total steam:

$$0.80X_1 + 0.6X_2 + 0.82X_3 + 0.65X_4 + 0.84X_5 + 0.76X_6 = 150$$

2. Max capacity of Boiler 1:

$$0.8X_1 + 0.82X_3 + 0.84X_5 \leq 100$$

3. Max capacity of Boiler 2:

$$0.6X_2 + 0.65X_4 + 0.76X_6 \leq 90$$

4. Gas contract:

$$X_5 + X_6 \geq 120$$

5. Oil supply limitation:

$$X_3 + X_4 \leq 20$$

6. Coal supply:

$$X_1 + X_2 \leq 150$$

Solve for variables using linear programming for minimum cost.

4
Reliability

I. INTRODUCTION

The reliability of a component or system can be represented in a statistical sense by the probability of a component or system performing satisfactorily at a particular time under a specified set of operating conditions. The definition of what constitutes 'satisfactory' may depend upon the nature of the system. Some devices, such as switches and valves, may have only an 'operate' or 'non-operate' mode. Other devices may be judged satisfactory or not depending on the required output level of some performance variable such as power or thrust. The present introductory discussion will consider the first two of the following four aspects of reliability:

1. The change in reliability of a component or system with age
2. The reliability of a system as influenced by the arrangement of components
3. The precision of estimates of reliability and other associated reliability parameters
4. The ability of a product to perform within specified limits under the influence of some external stress or environment

The object of the present discussion is to introduce some basic concepts and complete treatments can be found in the various references cited.

This major emphasis to date on the use of mathematical reliability models has been in the aero-space and defense industries. Particular attention has been given in the literature to studies of electronic systems. There has been somewhat less emphasis on the mathematical aspects of reliability as applied to mechanical systems. The reduced emphasis is not due to lack of interest, but rather to the comparatively high reliability of typical mechanical systems. In addition, high unit costs (of equipment for testing specimens) and the lengthy test requirements (because of good existing reliability) have limited the number of studies. Although complex electronic equipment is also costly, the components (tubes, transistors, resistors, etc.) are comparatively inexpensive and can be tested individually under a variety of controlled conditions. A study of the mathematical principles of reliability has many useful concepts to offer the designer, despite the lack of extensive quantitative design data on mechanical systems.

Attempts at establishing quantitative statements concerning reliability were initiated during World War II and were mainly concerned with developing good vacuum tubes and reliable radio communication.

Between 1945 and 1950 studies [4.29] revealed that:

1. A navy study made during maneuvers showed that the electronic equipment was operative only 30% of the time.
2. An army study revealed that between 2/3 and 3/4 of their equipment was out of commission or under repairs.
3. An Air Force study over a 5-year period disclosed that repair and maintenance costs were about 10 times the original cost.
4. A study uncovered the fact that for every tube in use there was one on the shelf and seven in transit.
5. Approximately one electronics technician was required for every 250 tubes.
6. In 1937 a destroyer had 60 tubes, by 1952 the number had risen to 3200 tubes.

It must be remembered that these studies were based on equipment produced during World War II by anyone able to walk to a production line. The engineering design work for many of these items was based on pre-World War II design concepts. The analytical techniques coming from design work on Korean and World War II weapon systems may have contributed as much to improvement of systems reliability as the mathematical concepts of reliability. Also note that airplanes in World War II were designed by using slide rules and desk calculators, where as the advent of analog and digital computers has allowed designers to simulate performance before committing themselves to a fixed design. In other words, more variations and parameters can be considered in the analyses today.

What the concept of reliability has done is to bring to engineering the benefits of statistics and probability for use in design. This in itself gives the engineer additional tools to use while designing.

Missile projects [4.29] such as the "Sparrow," "Regulus" and "Redstone" missiles and those since 1950 have used reliability concepts. The 10% reliability of the early Vanguard program increased to virtually 100% in the Minuteman. (The engineering design capabilities during this time period were also increasing by leaps and bounds.)

1. During the Korean War less than 30% of the combat airplane electronic equipment was operational. Later similar equipment is over 70% operational. (The Korean War was fought with almost half the planes and virtually all the ships of World War II Vintage.)
2. In 1958 only 28% of all United States satellite launchings were successful, whereas in 1962, 83% of all United States launchings were successful.
3. In 1959 passenger-car Warranties were for a period of 90 days or 4000 miles whichever came first. In 1997, 100,000 mile warranties are offered.

It is understood that Reliability is "the probability that a device will perform its specific function for a specific time under specific operating conditions." Note that to define Reliability

1. satisfactory performance must be stated
2. time is involved (either calendar time or number of operating cycles)
3. operating conditions must be stated
4. then after testing the probability can be estimated

There are several areas of interest in reliability for engineers:

1. designing with reliability in mind
2. measuring reliability
3. management or organization of systems for high reliability
4. prediction of reliability by means of mathematics

II. RELIABILITY FOR A GENERAL FAILURE CURVE

The best possible way to discuss Reliability would be to start with the basic part of its definition-probability.

There are various distribution curves for failure data which are not necessarily Gaussian. These alternative distributions are approximated by curve-fitting the failure data. Some of the distributions which can be found in handbooks [4.10] are:

Binomial distribution
Geometric distribution
Poisson distribution
Triangular distribution
Normal distribution

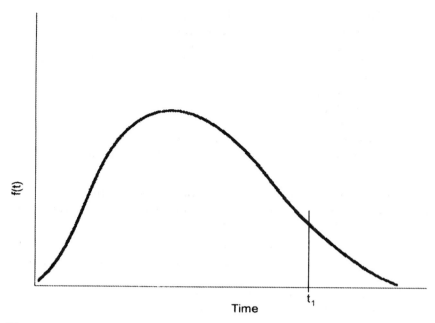

Figure 4.1. A failure curve.

  Log-normal distribution
  Gamma distribution
  Beta distribution
  Exponential distribution
  Weibull distribution

All of those listed above will not be covered in detail. They are only mentioned to show the various mathematical models which could be used. Those listed above are by no means inclusive. As always, a goodness-of-fit test should be conducted to determine whether the chosen distribution is appropriate.

Look at the normal curve where $\mu = 0$ and $\check{z} = 1$ from Eq. (1.1)

$$f(x) = \frac{1}{\sqrt{2\pi}} \exp\left[-\frac{x^2}{2}\right] \tag{4.1}$$

the data curve is

$$f(x) = \frac{1}{\check{z}\sqrt{2\pi}} \exp\left[-\frac{1}{2}\left(\frac{x-\mu}{\check{z}}\right)^2\right] \tag{4.2}$$

Note the variable x could just as well be the time variable t. Also the area under the curve is normalized, Eq. (4.1), knowing the whole area is $\sqrt{2\pi}$.

$$\int_{-\infty}^{+\infty} y\,dx = \int_{-\infty}^{\infty} \frac{1}{\sqrt{2\pi}} \exp\left[-\frac{x^2}{2}\right] dx \tag{4.3}$$

Therefore areas under portions of the curve can be interpreted as probabilities. Let the variable x be t, the integral

$$\int_{-\infty}^{+\infty} f(t)\,dt = area = A \tag{4.4}$$

Dividing by A

$$\int_{-\infty}^{\infty} \frac{f(t)}{A}\,dt = 1 \tag{4.5}$$

Therefore, any areas under the $f(t)$ versus t curve also represent probability and also represents the number of items tested if all failed during testing.

Reliability is the probability that a device will perform its specified function for a specified time under specified operating conditions.

Take a time t_1 on the $f(t)$ curve Fig. 4.1 and note that to the left of the t_1 line are the failures and to the right are the items which have not failed. In computing the reliability interest is in the percent of those which have not failed up to time t_1. Further, since the normalized area under the curve is 1, the area under the curve from t_1.

$$R(t_1) = \int_{t_1}^{\infty} \frac{f(t)}{A}\,dt \tag{4.6}$$

Another parameter, the Mean Time To Failure is useful.

$$MTTF = \int_{0}^{\infty} t\frac{f(t)}{A}\,dt = \int_{0}^{\infty} R(t)\,dt \tag{4.7}$$

III.   RELIABILITY FOR A RATE OF FAILURE CURVE

The concept of time as applied to mathematical models for reliability may refer to clock time (i.e. hours, minutes, etc.) or to the number of cycles of operation (e.g., number of times used, cycles of stress, etc.). For the purposes of the following discussion, it will be assumed that the conditions constituting failure have been defined. If a number N of identical items is tested for reliability until some N_f have failed, at some time t an empirical estimate of the reliability is

$$R(t) = \frac{N - N_f(t)}{N} = \frac{N_s(t)}{N} \qquad (4.8)$$

where N_s refers to the number of items remaining in service. Although tests are conducted on a limited sample, one would prefer to have N as large as possible in order to provide reasonable precision in the estimates computed from the data. The requirement for a large test sample is analogous to the conditions required for the experimental measurement of the probabilities associated with coin-flipping or dice-throwing. It is worth noting that, for games of chance, a reasonable mathematical model makes a priori predictions about the experimental results. In studying reliability, experiments should be conducted to infer a suitable mathematical model so that projections of future performances can be calculated.

The reliability, $R(t)$ is Eq. (4.8), or the probability of survival at time t. In a similar manner define unreliability or the probability of failure as

$$Q(t) = N_f(t)/N \qquad (4.9)$$

and note that

$$R(t) + Q(t) = 1.0 \qquad (4.10)$$

Assume that the variables $R(t)$ and $N_s(t)$ in the empirical definition as continuous (instead of discrete) in order to study reliability from a mathematical standpoint. Differentiating Eq. (4.10), dividing by N_s and substituting Eq. (4.9)

$$\frac{1}{N_s}\left[\frac{dR(t)}{dt} + \frac{dQ(t)}{dt}\right] = \frac{1}{N_s}\left[\frac{dR(t)}{dt} + \frac{dN_f(t)}{N dt}\right] = 0$$

and rearranging and multiply by N

$$0 = \frac{N}{N_s(t)}\frac{dR(t)}{dt} + \frac{dN_f(t)}{N_s(t)dt}$$

substituting Eq. (4.8)

$$0 = \frac{1}{R(t)}\frac{dR(t)}{dt} + \frac{dN_f(t)}{N_s(t)dt}$$

The second term is frequently called the instantaneous failure rate or hazard rate, $h(t)$ which yields

$$\frac{d[\ln R(t)]}{dt} + h(t) = 0 \qquad (4.11)$$

Integrating Eq. (4.11) and defining $R(0) = 1.0$

$$R(t) = e^{-\int_0^t h(\tau)d\tau} \tag{4.12}$$

Now consider the mathematical models for $h(t)$ in Appendix A and D.

IV. RELIABILITY FOR A CONSTANT RATE OF FAILURE CURVE

The form of Eq. (4.12) suggests considering $h(t)$ a constant as a simple failure model. This model is frequently called the exponential or constant hazard rate model. In addition to the obvious simplicity, there are sound physical reasons for seriously considering this model. Figure 4.2 shows a failure rate versus age (time) curve which is typical of the performance of many systems and some types of components.

The central portion of the curve Fig. 4.2 represents the useful life of the system and is characterized by chance or random failures. The high initial failure rate is due to shakedown or debugging failures and can be reduced by improving production quality control and/or breaking in equipment before leaving the factory. Aging failures are minimized by preventative maintenance–i.e. repair or replacement of parts susceptible to aging. Hence, in Fig. 4.2 there is a kind of empirical justification for assuming $h(t)$ a constant over a substantial portion of the life if a system, provided measurements are taken to minimize or eliminate the initial and wear out failures.

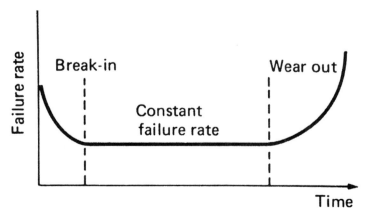

Figure 4.2. Typical bath tub aging curve.

As an example: telephone equipment for underwater Atlantic phone cables have been tested for a 20 years burn in so that the remaining 20 years life at lower constant rate failure is available.

From another standpoint, assume that chance or random events (failures) are most likely to cause unreliability. If these chance or random events have a small probability of occurrence in a large number of samples, the mathematical model might be described by a Poisson distribution.

$$P(n) = \frac{m^n \exp(-m)}{n!} \tag{4.13}$$

where m is the mean number of occurrences and $P(n)$ is the probability of an event occurring exactly n times. In reliability there is interest in the probability of no failures

$$R = P(0) = \frac{m^0 \exp(-m)}{0!} = \exp(-m) \tag{4.14}$$

The corresponding unreliability is represented by the series

$$Q = \sum_{n=1}^{\infty} P(n) = \sum_{n=1}^{\infty} \frac{m^n \exp(-m)}{n!} \tag{4.15}$$

Equations (4.14) and (4.15) satisfy the condition that

$$R + Q = 1.0 = P(0) + \sum_{n=1}^{\infty} \frac{m^n \exp[-m]}{n!} = \sum_{n=0}^{\infty} \frac{m^n \exp[-m]}{n!} \sum_{n=0}^{\infty} \exp[m - m]$$

$$\exp(m) = \frac{m0}{0!} + \frac{m^1}{1!} + \frac{m^2}{2!} + \frac{m^3}{3!} + \cdots = \sum_{n=0}^{\infty} \frac{m^n}{n!}$$

$$\tag{4.16}$$

Set $h(t) = \lambda$ and interpret λ as the failure rate and λt as the mean number of occurrences (in time t), hence

$$R(t) = \exp[-\lambda t] \tag{4.17}$$

A similar result is obtained by performing the integration indicated in Eq. (4.12), letting $h(t) = \lambda$. A continuous function $\lambda(t)$ is substituted for the discrete variable m for the purpose of developing a mathematical model.

The reciprocal of the failure rate, λ, is usually called the mean time to failure, MTTF, in a one-shot system. The exponential model is known as a one-parameter distribution because the reliability function is completely specified when the MTTF or $\frac{1}{\lambda}$ is known. Although the failure rate is constant, the failures are distributed exponentially with respect to time.

Equation (4.17) has been plotted in Fig. 4.3 to show how reliability is related to age and MTTF. The figure shows that:

1. Accurate MTTF estimates are necessary to define the reliability of relatively unreliable devices.
2. Age or operating time is an important parameter in determining the reliability of devices with poor reliability.
3. Accurate MTTF estimates are less important for highly-reliable devices.

It is also important to note that the reliability for the exponential model at $t = MTTF$ is only 0.368 (and not 0.5), and, the failure $Q(t) = 0.632$.

In complex equipment where wear out failure is significant, if the aging characteristics of different parts varies, then the failure pattern of the components as reflected in the reliability of the total system often appears as a series of random or chance events. *Hence, the reliability of a complicated system may appear to be an exponential function, even if the individual failure characteristics of the components are not of the exponential type.* Cumulative failure data for a marine power plant Fig.

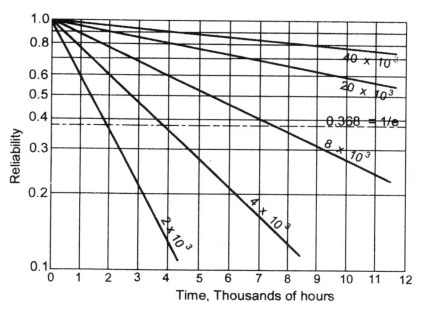

Figure 4.3. Constant reliability as a function of time.

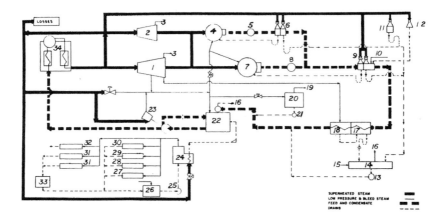

Figure 4.4.   Components for a typical marine power plant [4.30].

Table for Fig. 4.4 numbered components

1. Main turbine	17. Drum cleaner
2. Turbo generator	18. Low pressure heater
3. Gland leaks	19. To tanks
4. Aux. condensor	20. Flash distilling plant
5. Aux. condensor pump	21. Distilling heater drain pump
6. Aux. air ejector	22. Deaerating feed heater
7. Main condensor	23. Main feed pump
8. Main condensor pump	24. Contam steam generator
9. Main air ejector	25. Feed pump
10. Gland leak and vent	26. Drain tank
11. Distiller air ejector	27. Cargo dehumidation
12. Low pressure heater drum cleaner vent air ejector	28. Hot water
	29. Galley and laundry
13. Atmosphere drum tank drain pump	30. Ships heating
	31. Fuel oil tank heaters
14. Atmospheric drain tank	32. Steam atomizing
15. Make-up feed	33. Inspection tank
16. Vent	34. Boilers

4.4 are estimated along with some typical MTTF's by Harrington and
Riddick [4.8] and [4.30] in Table 4.1. In examining Fig. 4.4 component data,
keep in mind that there are 8760 hr/year (continuous operation) or about
2080 hr/year at 40 hr per week.

Table 4.1 Machinery plant component failure rates and mean time between failures [4.30]

Component	No. of units included	Total no. of failures	Total hours operation	MTBF	Failure per 100,000 hours	Reliability $R = e^{-\lambda t}$
Pumps-main feed	2	4	85,680	21,400	4.7	0.7542
Main condste.	2	4	80,600	20,150	5.0	0.7407
Aux. condste.	2	3	85,680	28,600	3.5	0.8104
Main circ.	2	8	80,600	10,080	10.0	0.5489
Aux. circ.	2	7	85,680	12,250	8.2	0.6113
Other SW	6	7	51,400	7,340	13.6	0.4421
Lube oil	2	3	80,600	26,900	3.7	0.8010
FO serv.	2	2	85,680	42,800	2.3	0.8711
FO trans.	2	0	13,700	13,700	7.3	0.6452
Main boiler	2					
Tubes		6	128,400	21,400	4.7	0.7542
Refractory		14	128,500	9,200	10.9	0.5200
SH tube supports		6	128,500	21,400	4.7	0.7542
Safety valves		13	128,500	9,900	10.1	0.5456
Soot blowers		17	18,500	7,560	13.2	0.4505
Drum desuperheater		3	128,500	42,900	2.3	0.8711
Superheat temp. control		5	128,500	25,700	3.9	0.7914
Feed reg. valve		9	128,500	14,300	7.0	0.5711
Generators	2	1	171,200	171,200	0.6	0.9646
Main turbines	2	0	161,200	161,200	0.6	0.9646
Main red. gear	1	1	80,600	80,600	1.2	0.9305
DFT	1	1	85,680	85,680	1.2	0.9305
HP feed heater	1	1	85,680	85,680	1.2	0.9305
LP feed heater	1	0	85,680	85,680	1.2	0.9305
FW evaporator	2	0	85,680	85,680	1.2	0.9305
Air ejector main	1	0	80,600	80,600	1.2	0.9305
Aux.	2	0	85,600	85,600	1.2	0.9305
Evap.	2	1	80,600	80,600	1.2	0.9305
Condensor-main	1	0	80,600	80,600	1.2	0.9305
Aux.	2	0	85,680	85,680	1.2	0.9305
Gas air heaters	2	5	128,500	25,700	3.9	0.7914
Forced draft blower	2	0	85,680	85,680	1.2	0.9305

*Based on operation for 6000 h.

EXAMPLE 4.1. The (hypothetical) data in Table 4.2 resulted from a reliability test. Plot a reliability curve and estimate the MTTF from the resulting straight line approximation. Compute the MTTF from the data and show the corresponding straight line approximation.

The notation is from Eq. (4.8) where:
N – number of samples in the test (24)
ΔN_f – number of samples which failed during the test interval time
N_S – average number of units still in service during the test interval
$R(t)$ – reliability at the end of the test interval.

Note $R(t)$ is known before the time interval starts or at the end, and the *true* location in the interval is never known. The $R(t)$ values are plotted here, some individuals plot points at the mid span of the interval (which is arbitrary). When the data is plotted (Fig. 4.5) note that

$$t = 0 \quad R(t) = 1 = \exp(-\lambda t)$$

$$t = (1/\lambda) \quad R(t) = \exp\left(-\lambda \frac{1}{\lambda}\right) = \exp(-1) = 0.368$$

The MTTF in this case is ≈ 4000 hr at the intersection of the best fitted line and $R(1/\lambda) = 0.368$.

The failure rate (failures per hour) is calculated from the first time interval

$$\lambda = \frac{\Delta N_f}{N_S} \frac{1}{\Delta t} = \frac{7}{\frac{1}{2}[24 + 17]} \frac{1}{1500 \text{ hr}} = 2.276 \times 10^{-4} \frac{\text{failures}}{\text{hr}}$$

The MTTF is a weighed function of the ΔN_f, the time interval and N set to 21 since 3 units did not fail

$$MTTF = \frac{1}{\lambda} = \frac{1}{21}\sum t_i \Delta N_{fi} = \frac{1}{21}[7(1.5) + 5(3.0) + 3(4.5) + 2(6.0)$$
$$+ 2(7.5) + 1(9.0) + 1(10.5)] \times 10^3 \text{ hr}$$

$$MTTF = \frac{1}{\lambda} = 4.0714 \times 10^3 \text{ hr} \quad (4000 \text{ hr from Fig. 4.5})$$

The algebra for the following calculation is not considered correct when calculating a $\bar{\lambda}$, because using an N of 24 instead of 21 the MTTF of 4148 hr is high compared to 4000 hr from Fig. 4.5

$$\bar{\lambda} = \frac{1}{21}[7(2.276) + 5(2.298) + 3(1.9048) + 2(1.667)$$

$$+ 2(2.222) + 1(1.481) + 1(1.905)] \times 10^{-4} = 2.1095 \times 10^{-4} \frac{\text{failures}}{\text{hr}}$$

The MTTF $= \frac{1}{\lambda} = \frac{1}{2.1095 \times 10^{-4}} = 4740 \text{ hr}$

Table 4.2 Test data and data reduction

Data $\times 10^3$ hr	0-1.5	1.5-3.0	3.0-4.5	4.5-6.0	6.0-7.5	7.5-9.0	9-10.5	24 in test
ΔN_f failures in interval Eq. (4.8) R(t)	7	5	3	2	2	1	1	total 21 failure 3 still running
	17/24 (0.7083)	12/24 (0.5000)	9/24 (0.3750)	7/24 (0.2917)	5/24 (0.2083)	4/24 (0.1667)	3/24 (0.1250)	
$N \dfrac{\Delta_f}{N_S}$	0.3415	0.3448	0.2857	0.2500	0.3333	0.2222	0.2857	
$\dfrac{\Delta N_f}{N_S} \dfrac{1}{\Delta t} = \lambda_i \times 10^{-4}$	2.276	2.298	1.9048	1.667	2.222	1.481	1.905	

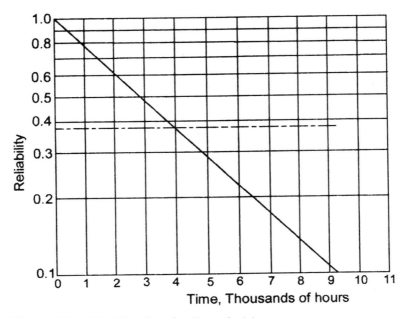

Figure 4.5. Reliability data for Example 4.1.

The convention is with $N = 21$

$$MTTF = \frac{1}{N}\sum\frac{N_{fi}}{\lambda_i} = \frac{1}{21}\left[\frac{7}{2.276} + \frac{5}{2.298} + \frac{3}{1.9048} + \frac{2}{1.667} + \frac{2}{2.222}\right.$$

$$\left. + \frac{1}{1.481} + \frac{1}{1.905}\right] \times 10^4 = 4822 \text{ hr}$$

It should be realized that $\lambda = h(\tau)$ in Eq. (4.12) and

$$R(t) = \exp(-\lambda t)$$

λ is the slope of the line Fig. 4.5 fitting the actual data by computer non-linear regression or the visual best fit. The selected line represents a smoothing of errors from the interval calculations. The general rule is to plot $R(t)$ and compare to constant failure rate, Gaussian, and Weibull reliability curves.

V. GAUSSIAN (NORMAL) FAILURE CURVE

The Gaussian or normal distribution function is sometimes used as the mathematical model for components or devices which fail primarily by wearing out. Equation (4.2) describes the two-parameter Gaussian model

$$R(t) = \frac{1}{\check{z}\sqrt{2\pi}} \int_t^\infty \exp\left[-\frac{1}{2}\left(\frac{t-\mu}{\check{z}}\right)^2\right] dt \tag{4.18}$$

where μ is the mean life of $\check{z}$ is the standard deviation, a measure of the dispersion of reliability values about the mean life. μ and $\check{z}$ are computed from a limited sample of experimental data. The discrete events (failures) are represented by a continuous model. The frequency or number of failures versus time is described by the familiar bell-shaped curve. The standard deviation is a measure to the peakedness of flatness of this distribution as illustrated in Fig. 4.6. The reliability as a function of time is the cumulative probability shown in Fig. 4.7.

Assuming that there are no censored observations estimate μ and $\check{z}$ from

$$\mu \approx \frac{\sum t_i}{N} \tag{4.19}$$

$$\check{z} \approx \sqrt{\frac{(\sum t_i - \mu)^2}{N-1}} = \sqrt{\frac{(\sum t_i^2) - \mu \sum t_i)^2}{N-1}} = \sqrt{\frac{N(\sum t_i^2) - N(\sum t_i)^2}{N(N-1)}} \tag{4.20}$$

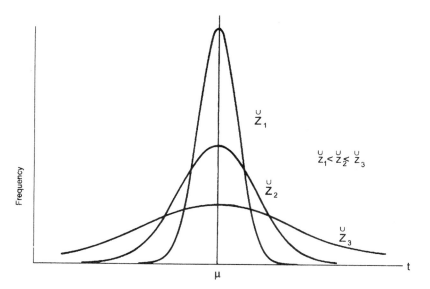

Figure 4.6. Normal distributions of failures with time.

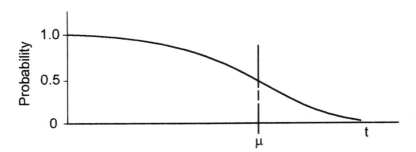

Figure 4.7.   Gaussian reliability curve with time.

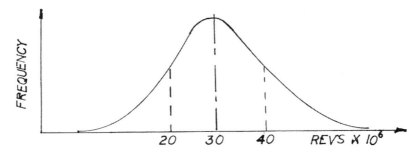

Figure 4.8.   Gaussian failure of a bearing Example 4.2.

Censored observations [1.7,1.22,4.24] will drop extreme values from the data set using statistical methods so that better values for μ and $\check{z}$ may be obtained with better confidence.

EXAMPLE 4.2.   Estimate the reliability of a bearing at 20×10^6 and at 40×10^6 if the mean life is 30×10^6 revolutions and the standard deviation is 5×10^6 Revs, in Fig. 4.8.

The simplest method of solution is to use tabulated values of the probability integral. Most tables are normalized with the argument given in terms of the number of standard deviations (i.e. t/μ). In this case, there is interest in computing the reliability of $\pm 2\,\check{z}$ either side of the mean life. Graphically speaking, the area from 20×10^6 cycles to 40×10^6 cycles. Since the area Eq. (4.2) under the normalized distribution curve $\pm 2\,\check{z}$ is 0.9544

$$R(20 \times 10^6) = 0.5 + \frac{1}{2} \int\limits_{-2\check{z}+\mu}^{2\check{z}+\mu} \frac{1}{\check{z}\sqrt{2\pi}} \exp\left[-\frac{1}{2}\left[\frac{x-\mu}{\check{z}}\right]^2\right] dx$$

from Fig. 4.8 and a math handbook.

$$R(20 \times 10^6) = 0.5 + 0.3413 + 0.1359 = 0.9772$$

Similarly,

$$R(40 \times 10^6) = 0.5 - 0.3413 - 0.1359 = 0.0228$$

VI. CONFIGURATION EFFECTS ON RELIABILITY

A. Series System

Components in series are frequently represented by a block diagram Fig. 4.9.

The system composed of elements A, B, and C represents a series of machines or operations which must be performed (or operate) in unbroken sequence (or simultaneously) to achieve the required output. *Since all elements must operate, it is the mathematical probability.*

$$R(\text{system}) = P(\text{system}) = P(A) \text{ and } P(B) \text{ and } P(C)$$

If the probabilities are independent,

$$R(\text{systems}) = P(A)P(B)P(C) = R(A)R(B)R(C) = R(ABC) \qquad (4.21)$$

This relationship is analogous to the more familiar result for efficiencies, where the efficiency of a machine is obtained as a product of the efficiencies for the parts. The series Christmas tree lights represent this when one light burns out all the lights fail or go out. The system fails and one can't easily find the light that burned out but one knows the system failed when one or more lights burns out. It can also be noted for the exponential model of Eq. (4.17) that one simply sums the exponents in order to obtain the reliability for a series system.

B. Parallel System

Components in parallel are represented by a block diagram Fig. (4.10). where the desired output is obtained if any one of the elements A, B, or C operates successfully.

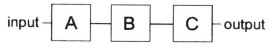

Figure 4.9. Series reliability block diagram.

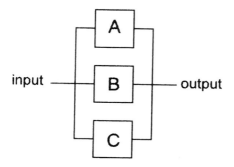

Figure 4.10. Parallel reliability block diagram.

The probability that A and B and C will all *fail* to work is

$$Q(\text{system}) = Q(A)Q(B) \text{ and } Q(C)$$
$$= (1.0\text{-}R(A))(1.0\text{-}R(B))(1.0\text{-}R(C)) \tag{4.22}$$

Hence resulting in a form of Eq. (2.3)

$$R(\text{system}) = 1.0\text{-}Q(\text{system}) = R[A + B + C]$$
$$= 1.0\text{-}[1.0\text{-}R(A)][1.0\text{-}R(B)][1.0\text{-}R(C)] \tag{4.23}$$

EQUATION 4.22 APPLIED IF ONLY ONE ELEMENT OPERATES. The extra elements are termed redundant. They are necessary only in the event of failure in the primary element. As an example, the parallel office fluorescent lights when one burns out the rest give light and the system has not failed. The burnt fluorescent can be replaced quickly and the system does not stop functioning.

C. Series–Parallel Systems

Systems with groups of components in parallel and others in series can usually be analyzed by applying Eqs (4.21)–(4.23) to parts of the system and then reapplying the equations to groups of parts. The process is analogous to the calculation of resistance (or conductance) in complex electrical circuits by repeated application of the simple rules for series and parallel circuits.

EXAMPLE 4.3. In order to simplify the reliability block diagram in Fig. 4.11:

Compute reliability for A1, A2, A3 in simple or partial parallel $= A'$
Compute reliability for A4, A5, A6 in simple or partial parallel $= A''$

Compute reliability for B1, B2 in series = B′
Compute reliability for B3, B4 in series = B″
Compute reliability for B5, B6 in series = B‴
Compute reliability for B7, B8 in series = B^{IV}

Then shown in Fig. 4.12
The further reduction for Fig. 4.12 requires one to:

Compute reliability for B′ and B″ in parallel = B(1)
Compute reliability for B‴ and B^{IV} in parallel = B(2)

The results in Fig. 4.13
Figure 4.13 can be simplified further to:

Compute reliability for A′, B(1), C1 in series = A(1)
Compute reliability for A″, B(2), C2 in series = A(2)

Then in Fig. 4.14 the results are

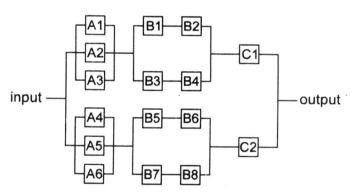

Figure 4.11. Complex reliability block diagram.

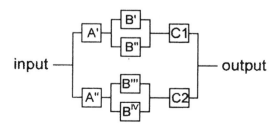

Figure 4.12. Figure 4.11 simplified block diagram.

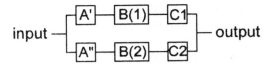

Figure 4.13. Figure 4.11 block diagram further simplified.

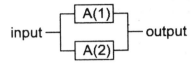

Figure 4.14. Final simplification for Fig. 4.11 block diagram.

Compute reliability from Fig. 4.14 for A(1), A(2) in simple partial parallel from Eq. (4.23). Reliability of components in series and parallel with constant rates of failure with constant rate of failure are treated.

D. Reliability of Series Components

The reliability of the series components is

$$R_S = R_1 \cdot R_2 \cdot R_3 \cdots R_n = \prod_{i=1}^{n} R_i \tag{4.24}$$

with a constant rate of failure

$$R_S = \exp[-\lambda_1 t] \exp[-\lambda_2 t] \dots \exp[-\lambda_n t]$$
$$R_S = \exp[-\Sigma \lambda_i t] \tag{4.25]}$$

for the above expression to hold:

1. The system reliability configuration must truly be a series one
2. The reliabilities of the components must be independent
3. The components must be governed by a constant-hazard rate model

The MTTF is

$$MTTF = \frac{1}{\sum\limits_{i=1}^{n} \lambda_n} = \frac{1}{\lambda_1 + \lambda_2 + \cdots + \lambda_n} \tag{4.26}$$

E. Reliability of Parallel Components

The reliability of parallel components is

$$R_p = 1 - \left[\prod_{i=1}^{i=n}(1 - \exp[-\lambda_i t]) \right] \tag{4.27}$$

for two components in parallel with different failure rates

$$R_p = 1 - (1 - \exp[-\lambda_1 t])(1 - \exp[-\lambda_2 t])$$
$$= \exp[-\lambda_1 t] + \exp[-\lambda_2 t] - \exp[-\lambda_1 + \lambda_2)t]$$

$$MTTF = \int_0^\infty R_p dt \tag{4.28}$$

$$= \left[-\frac{\exp[-\lambda_1 t]}{\lambda_1} - \frac{\exp[-\lambda_2 t]}{\lambda_2} + \frac{\exp[-(\lambda_1 + \lambda_2)t]}{(\lambda_1 + \lambda_2)} \right] \Big|_0^\infty$$

Two parallel components

$$MTTF = \frac{1}{\lambda_1} + \frac{1}{\lambda_2} - \frac{1}{\lambda_1 + \lambda_2} \tag{4.29}$$

Two and more parallel components can be developed in the same derivation. For three parallel components of different failure rates [4.8],

$$MTTF \text{ or } MTBF = \frac{1}{\lambda_1} + \frac{1}{\lambda_2} + \frac{1}{\lambda_3} - \frac{1}{(\lambda_1 + \lambda_2)} - \frac{1}{(\lambda_2 + \lambda_3)}$$
$$- \frac{1}{(\lambda_2 + \lambda_3)} + \frac{1}{(\lambda_1 + \lambda_2 + \lambda_3)} \tag{4.30}$$

When the rates are equal $\lambda_1 = \lambda_2$ for two components [4.29]

$$MTBF = \frac{2}{\lambda} - \frac{1}{2\lambda} = \frac{3}{2\lambda} \tag{4.31}$$

The failure rates equal for 3 parallel components

$$MTBF = \frac{3}{\lambda} - \frac{3}{2\lambda} + \frac{1}{3\lambda} = \frac{1}{\lambda}\left[3 - \frac{3}{2} + \frac{1}{3} \right] = \frac{11}{6\lambda} \tag{4.32}$$

The constant failure rates [4.8,4.20] for two to five parallel components yields the following reliabilities.

Two parallel components

$$R_p = 2\exp[-\lambda t] - \exp[-2\lambda t] \tag{4.33}$$

Three parallel components

$$R_p = 3\exp[-\lambda t] - 3\exp[-2\lambda t] + \exp[-3\lambda t] \tag{4.34}$$

Four parallel components

$$R_p = 4\exp[-\lambda t] - 6\exp[-2\lambda t] + 4\exp[-3\lambda t] - \exp[-4\lambda t] \qquad (4.35)$$

Five parallel components

$$R_p = 5\exp[-\lambda t] - 10\exp[-2\lambda t] + 10\exp[-3\lambda t] - 5\exp[-4\lambda t] \\ + \exp[-5\lambda t] \qquad (4.36)$$

F. Reliability of Standby Components [4.24]

The standby unit (Fig. 4.15) is in parallel with a primary unit, however, the standby is switched on only when the primary unit fails. The λ rates are the same for both units and all standbys.

The Poissons distribution yields an identity which applies

$$\left[1 + \frac{\lambda t}{1!} + \frac{(\lambda t)^2}{2!} + \cdots + \frac{(\lambda t)^n}{n!}\right]\exp[-\lambda t] = 1 \qquad (4.37)$$

When $n = 1$ (one standby)

$$R = [1 + \lambda t]\exp[-\lambda t] \qquad (4.38)$$

$n = 2$ (two standbys with a switch to primary)

$$R = \left[1 + \frac{\lambda t}{1!} + \frac{(\lambda t)^2}{2!}\right]\exp[-\lambda t] \qquad (4.39)$$

for n units as standbys

$$R = \left[1 + \frac{\lambda t}{1!} + \frac{(\lambda t)^2}{2!} + \cdots + \frac{(\lambda t)^n}{n!}\right]\exp[-\lambda t] \qquad (4.40)$$

EXAMPLE 4.4. A water pump station Fig. 4.16 has been set up for high reliability.

1. Calculator $R(t)$ for the system

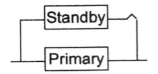

Figure 4.15. Standby systems.

2. Select typical failure rates find $t_{overhaul}$ for pumpset 1 or 2 when system $R(t) = 0.95$ What is the motor pump set reliability at this time?

3. First draw a reliability block diagram for one system delivering water pressure and add the standbys after the model is developed.

The numbered components for Figs. 4.16–4.18 are

1. Electric drive pump
2. Valve
3. Electric power
4. Pressure flow regulators
5. Electric power standby
6. Motor pump set 1 standby

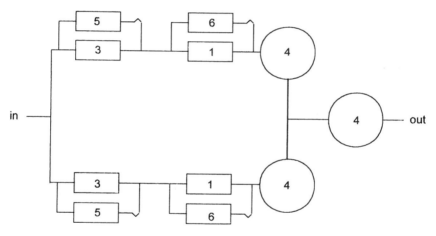

Figure 4.16. Water pump station.

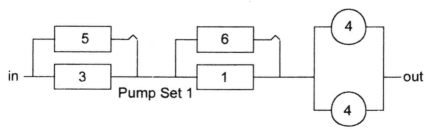

Figure 4.17. Half of water pump station Fig. 4.16.

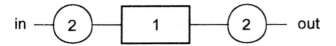

Figure 4.18. Motor pump set 1 for Fig. 4.17.

The λs are calculated using Appendix D with $K_F = 10$ in Eq. (D.2) and λ_G values from Table D.3. The λ_Gs are stated for failures 10^{-6} h

[upper extreme, mean, lower extreme] $\times 10^{-6}$

Electric drive pump
λ_G, Pump [27.4,13.5,2.9] with $\lambda_1 = \lambda_G K_F$
Shut off valves
λ_G, valves [10.2, 6.5, 1.98] with $\lambda_2 = \lambda_G K_F$
Electric power
λ_G, generator [2.41, 0.9, 0.04] with $\lambda_3 = \lambda_G K_F$
Pressure regulators
λ_G, flow pressure regulars [5.4, 2.14, 0.70] with $\lambda_4 = \lambda_G K_F$

In the motor pump set 1 (Fig. 4.18) reliability, use the upper extreme in Eqs. (4.24) and (4.25)

$$R_{1,2} = R_1 R_2 R_2 = \exp[-\Sigma \lambda_i t] = \exp[-(\lambda_1 + 2\lambda_2)t] = \exp\left[-\frac{478}{10^6} t\right]$$

Electric Power

$$R_3 = \exp[-\lambda_3 t]$$

Pressure flow regulators in parallel for equal λs Eq. (4.33) is Fig. 4.17

$$R_4 = 2\exp[-\lambda_4 t] - \exp[-2\lambda_4 t]$$

with one stand by pump and electric power, reliability increases by $(1 + \lambda_i t)$ Eq. (4.38) the system reliability is in series Eq. (4.21) and (4.38)

$$R_{\text{system}} = \{R_{12}(1 + \lambda_{12}t)\}\{R_3(1 + \lambda_3 t)\}\{R_4\}$$

R_{system} for top and bottom loops are the same then in parallel Fig. 4.16 from Eq. (4.23)

$$R_{\text{system}} = 1 - (1 - R_{\text{s top}})(1 - R_{\text{s bottom}})$$

Let's evaluate R_{system} using $t = 8760$ hr or one year.

A. Evaluate $R_{1,2}$ with standbys to the motor pump set 1 Eq. (4.38)

$$R_{1,2} = \left[1 + \frac{478}{10^6}(8760)\right]\exp\left[-\frac{478}{10^6}(8760)\right] = 0.0788$$

Note the poor reliability with $K_F = 10$ now change λ_G from 27.4×10^{-6} to 2.9×10^{-6} on the pump and the shut off value λ_G from 10.2×10^{-6} to 1.98×10^{-6} will change $R_{1,2}$.

Calculate $R_{1,2}$, again

$$R_{1,2} = \left[1 + \frac{68.6}{10^6}(8760)\right] \exp\left[-\frac{68.6}{10^6}(8760)\right] = 0.8778$$

much better but still not great. Use 68.6×10^{-6} for $\lambda_{1,2}$

B. Evaluate electric power with standby

$$R_3 = \left[1 + \frac{24.1}{10^6}(8760)\right] \exp\left[-\frac{24.1}{10^6}(8760)\right] = 0.9806$$

C. Pressure flow regulators for a parallel setup (Eq. (4.33))

$$R_4 = 2\exp\left[-\frac{54}{10^6}(8760)\right] - \exp\left[-2\frac{54}{10^6}(8760)\right] = 0.8580$$

Need low extreme for less failures, and increased reliability. Lets evaluate for top loop and then top and bottom loops in parallel

Top Loop $\quad R_{\text{system}} = \{0.8778\}\{0.9806\}\{0.8580\} = 0.7385$

Parallel $\quad R_{\text{system}} = 1 - (1 - 0.7385)(1 - 0.7385)$
$$= 1 - 0.06838 = 0.9316$$

Need low extremes in all components, hence, increased reliability. Motor pump set 1

$$R_{1,2} = \exp\left[-\frac{68.6}{10^6}(8760)\right]$$
$$R_{1,2} = 0.5483$$

This is not good! However, with standby yields 0.8778.

Redo B For electric power with low extreme $\lambda = (0.04 \times 10^{-6}) \times 10 = 0.4 \times 10^{-6}/\text{hr}$ using Eq. (4.38)

$$R_3 = \left[1 + \frac{0.4}{10^6}(8760)\right] \exp\left[-\frac{0.4}{10^6}(8760)\right] = (1.0035)(0.9965) = 1$$

Redo *C* For pressure regulators with lower extreme $\lambda = (0.70 \times 10^{-6}) \times 10 = 7 \times 10^{-6}/\text{hr}$. Eq. (4.33)

$$R_4 = 2 \exp\left[-\frac{7}{10^6}(8760)\right] - \exp\left[-2\left(\frac{7}{10^6}\right)(8760)\right]$$
$$= 1.8810 - 0.8846$$
$$R_4 = 0.9965$$

$R_{\text{system}} = \{0.8778\} \{1.00\} \{0.9965\} = 0.8747$ for the top loop

Now since the system is made of two parallel components, Eq. (4.23)

$$R_{\text{system}} = 1 - (1 - R_{\text{sys}})(1 - R_{\text{sys}})$$
$$= 1 - (1 - 0.8747)(1 - 0.8747)$$
$$R_{\text{system}} = 0.9843 \quad \text{yearly overhaul for pump needed for the parallel}$$
$$\text{setup}$$
$$Q_{\text{system}} = \left(\frac{1.57}{100}\right) \text{failures}$$

EXAMPLE 4.5. Determine the reliability of the automotive gear box Fig. 4.19 noting 3rd gear is used 93% of the time; 2nd gear 3%, 1st gear 3% and reverse 1%. Find the time to reduce the reliability to 0.90. The reliability of 3rd, 2nd, 1st, and reverse are each series in components and the operation of the gear box is a series combination of 3rd, 2nd, 1st and reverse.

The Reliability Model for 3rd Gear

Assuming the driver wishes to operate the car in 3rd gear, maximum speed, shifts F into the position shown and pushed D to the left Fig. 4.19 so that the clutch piece C engages with B, in which case P runs at the same speed as the engine shaft E. This means a series reliability model Eq. 4.24 for 93% of the time shown in Fig. 4.20. In Fig. 4.20 the numbers represent reliabilities

R_1 – A bearing and seal
R_2 – A bearing and seal
R_3 – A jaw clutch
R_4 – Shifting fork
R_5 – Left shaft
R_6 – Right shaft
R_7 – Housing

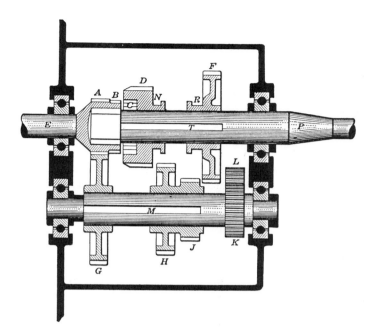

Figure 4.19. Automobile gear box used with permission [4.1]

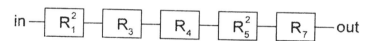

Figure 4.20. Third gear series reliability model for Fig. 4.19.

Now for 93% of the time with third gears or direct drive.

$$R_{3rd} = R_1^2 R_3 R_4 R_5^2 R_7$$

Second Gear Reliability Model

Second highest, 2nd gear, speed is obtained Fig. 4.19 by slipping D to the right until it comes into contact with H, the ratio of gears then being A to G and H to D; F remains as shown. This creates a series reliability model Eq. (4.24) for 3% of the time, shown in Fig. 4.21, resulting in the numbered reliabilities are

R_1 – four bearings and seals
R_2 – Two gear pairs

R_3 – Lower shaft
R_4 – Left shaft
R_5 – Right shaft
R_6 – Shifting fork
R_7 – Housing

$$R_{3rd} = R_1^4 R_2^2 R_3^3 R_6 R_7$$

First Gear Reliability Model

The same reliability model as 2nd gear for 3% of the time in Fig. 4.21.

For lowest speed, first gear, D is placed as shown in Fig. 4.19 and F slid into contact with J

$$R_{1st} = R_{2nd}$$

Reverse Reliability Model

The same reliability model as 1st and 2nd gear except adding 5–6 bearings (use 6), 3 gear pairs, and 4 total shafts. The shaft P Fig. 4.19 and the car are reversed by moving F to the right until it meshes with L, the gear ratio being A to G and K to L to F. Note: K gear is in front of gear L. The reliability Fig. 4.22 is shown and the numbers represent

R_1 – Six bearing and seals
R_2 – Three gear pairs
R_3 – Four shafts
R_4 – Shifting fork
R_5 – Housing

The reliability for one percent of the time is in reverse as follows

$$R_{\text{reverse}} = R_1^6 R_2^3 R_3^4 R_4 R_5$$

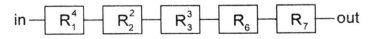

Figure 4.21. Second gear series reliability model for Fig. 4.19.

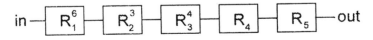

Figure 4.22. Reverse gear series reliability model for Fig. 4.19.

The reliability of the components are from Table D.3. The λs for the components are stated (high extreme) (mean) (low extreme) $\times 10^{-6}$ for failure rates/hr.

Bearings and Seals

From Eq. (D.2)

$$\lambda = \lambda_G K_F$$

$K_f = 30$ for rail-mounted equipment with Eq. (D.2.), Table D.2

    (a)   Ball bearing high speed heavy duty (high, mean, low) $\times 10^{-6}$ failure/hr

$$\lambda_G = (3.53, 1.8, 0.072) = \lambda_{G1}$$

    (b)   Rotating seals

$$\lambda_G = (1.12, 0.7, 0.25) = \lambda_{G2}$$

For all -6 components use the high extreme 3.53×10^{-6} for ball bearing and 1.12×10^{-6} for rotating seals which means shorter life, cheaper parts, and maybe a good design. For a bearing and its seals reliability

$$R = \exp[-K_F(\lambda_{G1} + \lambda_{G2})t]$$
$$= \exp\left[-\frac{139.5}{10^6 \text{ hr}}t\right]$$

Gear Pairs

$K_F = 30$

    (c)   Spur gears for high failure rate

$$\lambda_G = (4.3, 2.175, 0.087) = \lambda_{G3}$$

$$R_{\text{gear}} = \exp[K_F \lambda_{G3} t] = \exp\left[-\frac{129}{10^6 \text{ hr}}t\right]$$

Shafting

$K_F = 30$

    (d)   Shafting

$$\lambda_G = (0.62, 0.35, 0.15) = \lambda_{G4}$$

$$R_{\text{shafting}} = \exp[K_F \lambda_{G4}] = \exp\left[-\frac{18.6}{10^6 \text{ hr}}t\right]$$

Shifting Fork (Fig. 4.23)

$K_F = 30$

    (e)   Three mechanical joints

$$\lambda_G = (1.96, 0.02, 0.011) = \lambda_{G5}$$

    (f)   Three structural sections

$$\lambda_G = (1.35, 1.0, 0.33) = \lambda_{G6}$$

Combining Fig. 4.23 there are three mechanical joints and structural member in series.

$$R_{\text{shift}} = R_{\text{joints}} R_{\text{struct}} = \exp[-3K_F(\lambda_{G5} + \lambda_{G6})t]$$
$$= \exp\left[-\frac{297.9}{10^6 \text{ hr}}t\right]$$

Housing

$K_F = 30$

    (g)   housing, cast, machined bearing surfaces

$$\lambda_G = (0.91, 0.40, 0.016) = \lambda_{G7}$$

$$R_{\text{housing}} = \exp[-K_F\lambda_{G7}t] = \exp\left[-\frac{27.3}{10^6 \text{ hr}}t\right]$$

Jaw Clutch

$K_F = 30$

    (h)   Jaw clutch

$$\lambda_G = (1.1, 0.04, 0.06) = \lambda_{G8}$$

Note: 0.04 can not be the average. Should be 0.08 or 0.58 per an error in data printout.

$$R_{\text{jaw}} = \exp[-K_F\lambda_{G8}t] = \exp\left[-\frac{33}{10^6 \text{ hr}}t\right]$$

The automobile gear box functions with 1st, 2nd, 3rd, and reverse and the

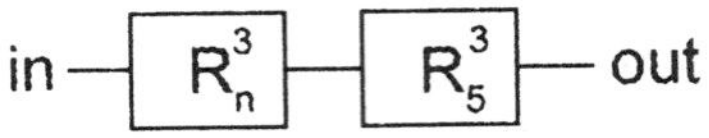

in —| R_n^3 |—| R_5^3 |— out

Figure 4.23. Shifting fork reliability model for Fig. 4.19.

time factor is

$$t_{3rd} = 0.93t \qquad t \text{ is the actual gear box operating time}$$
$$t_{2nd} = 0.03t$$
$$t_{1st} = 0.03t$$
$$t_{reverse} = \frac{0.01t}{1.00t}$$

The terms are substituted into the reliability equation to obtain the system reliability developed in Table 4.3.

$$R_{system} = (R_{3rd})(R_{2nd}(R_{1st})(R_{reverse})$$

$$R_{system} = \exp\left[-\frac{715.248}{10^6 \text{ hr}}t\right]$$

Let $R_{system} = 0.90$ and take $\ln_e$ of both sides

$$-\frac{715.248}{10^6 \text{ hr}}t = -0.1054$$

$t = 147.3$ hr continuous operation at rated power
Should lower extreme failure values be used, the hours would increase.
Note: Alone for 3rd gear

$$R_{3rd} = \exp\left[-\frac{627.19}{10^6 \text{ hr}}147.3 \text{ hr}\right] = 0.918$$

Lower failure values for just 3rd gear components

$2(\lambda_{G1} + \lambda_{G2}) = 2(0.072 + 0.25) = 0.644$ not 9.3 (a) (b) bearings and seals
$\lambda_{G8} = \qquad\qquad\qquad\qquad\qquad\quad = 0.060$ not 1.1 (h) jaw clutch
$\lambda_{G3} = \qquad\qquad\qquad\qquad\qquad\quad = 0.087$ not 4.3 (c) spur gears
$2\lambda_{G4} = \qquad\qquad\qquad\qquad\qquad\quad = 0.300$ not 1.24 (d) shafting

Table 4.3 Summary of transmission λs and a check on λ sums

Components	3rd	2nd	1st	Reverse	$\Sigma\lambda$s
Bearing seals	2(0.93)(139.5)	4(0.03)(139.5)	4(0.03)(139.5)	6(0.01)(139.5)	301.32
Jaw clutch	(0.93)(33)	None	None	None	30.69
Gear parts	None	2(0.03)(129)	2(0.03)(129)	3(0.01)(129)	19.35
Shafts	2(0.93)(18.6)	3(0.03)(18.6)	3(0.03)(18.6)	4(0.01)(18.6)	38.69
Shifting fork	(0.93)(297.9)	(0.03)(297.9)	(0.03)(297.9)	(0.01)(297.9)	297.9
Housing	(0.93)(27.3)	(0.03)(27.3)	(0.03)(27.3)	(0.01)(27.3)	27.3
λsums	627.192	35.91	35.91	16.236	715.248

$3(\lambda_{G5} + \lambda_{G6}) = 3(0.011 + 0.33) = 1.023$ not 9.93 (e) shifting fork

$\lambda_{G7} = \qquad\qquad = 0.016$ not 0.91 (g) cast housing

$\Sigma\lambda i = \qquad$ total slum $\qquad = 2.113$ not 26.78

λ_{3rd} reduces from $627.192 \left(\dfrac{2.13}{26.78}\right)$ to 49.885

the system λ without corrections to rest of columns in Table 4.3 is

$$\lambda = 715.248 - 627.192 + 49.885 = 137.94$$

When $R_{\text{system}} = 0.90$

$$-\frac{137.94}{10^6}t = -0.1054$$

$t = 764.1$ hour increase of 5.19 to the prior value. The change in the rest of the gear combinations would increase the operating hours.

REFERENCES

4.1.   Angus RW. The Theory of Machines, New York: McGraw-Hill Inc, 1917.

4.2.   Bain LJ. Statistical Analysis of Reliability and Life-Testing Models (Theory and Methods), Marcel Dekker, 1978.

4.3.   Bazovsky I. Reliability Theory and Practice, Englewood Cliffs, NJ: Prentice-Hall, 1965.

4.4.   Benjamin JA, Cornell CA. Probability, Statistics and Decisions for Civil Engineers, New York: McGraw-Hill Book Co, 1970.

4.5.   Bompas-Smith JH. Mechanical Survival, London: McGraw-Hill, 1973.

4.6.   Calabro SR. Reliability Principles and Practices, New York: McGraw Hill Inc, 1962.

4.7.   Hahn GJ, Shapiro SS. Statistical Models in Engineering, New York: John Wiley and Sons, 1967.

4.8.   Harrington RL, Riddick Jr RP. Reliability Engineering Applied to the Marine Industry, Vol. 1, Marine Technology, 1964.

4.9.   Ireson WG. Reliability Handbook, New York: McGraw-Hill Inc, 1966.

4.10. Kecicioglu D. Reliability Engineering Handbook, Vol. I. and II, Englewood Cliff, NJ: Prentice-Hall, 1991.

4.11. Lloyd DK, Lipon M. Reliability, Management Methods and Mathematics, Englewood Cliffs, NJ: Prentice-Hall, 1977.

4.12. King JR. Probability Charts for Decision Making, Industrial Press, 1971.

4.13. Mann NR, Schafer RE, Sing Purwalla ND. Methods for Statistical Analysis of Reliability and Life Data, New York: John Wiley and Sons, 1974.

4.14. Mechanical Reliability Concepts ASME, 1965.

4.15. Nelson W. Accelerated Testing, New York: John Wiley and Sons, 1990.

4.16. Pieruschka E. Principles of Reliability, Englewood Cliffs, NJ: Prentice-Hall, 1963.

4.17. RADC Non-Electronic Reliability Note Book RADC-TR-85-194 DTIC Alexandria, VA.

4.18. Non-Electronic Parts reliability Data, NPRD 95, Reliability Analysis Center, Rome NY, 1995.

4.19. RAC Journal, Reliability Analysis Center, Rome NY.

4.20. Reliability Handbook, Navy Ships 94501, August 1968.

4.21. Rothbart HA. Mechanical Design and Systems Handbook, New York: McGraw-Hill Book Co, 1964.

4.22. Smith DJ. Reliability Engineering, Barnes and Noble, 1972.

4.23. Shooman M. Probabilistic Reliability an Engineering Approach, New York: McGraw-Hill Inc, 1968.

4.24. Vidosic JP. Elements of Design Engineering, New York: The Ronald Press Co, 1969.

4.25. Von Aluen WH. (ed) Reliability Engineering, Englewood Cliffs, NJ: Prentice-Hall, 1964.

4.26. Wiesenberg RJ. Reliability and Life Testing of Automotive Radiators, General Motors Engineering Journal, 3rd Quarter 1962.

4.27. Woods BM, Degarmo ED. Introduction to Engineering Economics MacMillan Co, 1942.

4.28. Woodward III JB. Reliability Theory in Marine Engineering, Society of Naval Architects and Marine Engineers, Cleveland, 1 February 1963.

4.29. What is Reliability Engineering? Product Engineering, 16 May, 1960.

4.30. Riddick Jr RP. Application of Reliability Engineering to the Integrated Steam Power Plant, Proceedings on Advance Marine Engineering Concepts for Increase Reliability, University of Michigan, February 1963.

PROBLEMS

PROBLEM 4.1

A manufacturer sells a motor with major components having the following reliability characteristics:

  (I)   Electrical failure (insulation, windings, etc.) $MTTF = 20,000$ hr
  (II) Mechanical failure (impeller, casing, etc.) $MTTF = 10,000$ hr
  (III) Bearing wearout (2 bearings) $\mu = 1800$ hr $\sigma = 600$ hr, each
  (IV) Brush wearout (2 brushes) $\mu = 1000$ hr $\sigma = 200$ hr, each
  (a)   Calculate the reliability at 500 hr
  (b)   If the manufacturer has a 500 hr guarantee, how many motors will he have to replace or repair per 1000 sold?

PROBLEM 4.2

The motor in problem 1 is improved by using sealed bearings and by using a better quality alloy in the casing and impeller. The improved motor has the following reliability characteristics:

Electrical failure (insulation, windings, etc.) $MTTF = 20,000$ hr
Mechanical failure (impeller, casing, etc.) $MTTF = 20,000$ hr
Bearing wearout (2 brngs) $\mu = 2500$ hr $\quad \sigma = 600$ hr, each
Brush wearout (2 brushes) $\mu = 1000$ hr $\quad \sigma = 200$ hr, each

(a) What is the reliability of the improved model at 500 hr
(b) If the manufacturer wishes to have a replacement or repair rate of 5 motors per 100 sold, what should be his guarantee period?

PROBLEM 4.3

The following data were obtained from the reliability testing of a group of special gear boxes:

Hours $\times 10^4$	0–1	1–2	2–3	3–4	4–5	5–6	6–7	7–8	
Numbers failed	21	12	6	3	1	1	0	1	total $= 45$

Find the MTTF by plotting the data on semi-log paper.

PROBLEM 4.4

The following data were obtained from tests on hydraulic valves:

Hours cycles $\times 10^3$	0–1	1–2	2–3	3–4	4–5	5–6	
Numbers failed	11	6	3	2	1	1	total $= 24$

Find the MTTF by plotting the data on semi-log paper.

PROBLEM 4.5

The main boiler feed pump in a power plant has a $MTTF = 200,000$ hr.

(a) Find the reliability after one year of continuous operation at 24 hr/day, 7 days/week.
(b) Find the reliability after one year of operation at 40 hr/week.

PROBLEM 4.6

A propulsion system with four boilers, a two propeller outputs with shafting, reduction gears, and two turbines have steam supplied with two arrangements:

    (a)   One-boiler MTTF of 350,000 hr providing half-speed with three boilers on standby.

    (b)   Two boilers in series providing cruise speed with the other two in series on standby.

Find the reliability of both conditions and the over haul time if $R(t) = 0.95$ as a criterion. Find the reliability in cruise speed if remaining boilers are not in series standby but are used separately as standbys on the individual boilers in operation.

PROBLEM 4.7

The cross section shown in Problem 4.7 gives an indication of parts in a hand held power saw. Find the reliability of the saw with the information in Appendix D. Roughly estimate the time for the reliability to equal 0.90. [Note: This is a crude estimate.]

PROBLEM 4.8

Variable speed pulley patent drawings Problem 4.8 shows the pulley in two extreme positions. Estimate the reliability from information in Appendix D and find the time when $R(t) = 0.90$.

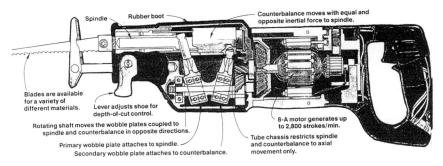

Prob 4.7. Super Sawall cross section (Permission of Milwaukee Electric Tool Corp. Brookfield, WI)

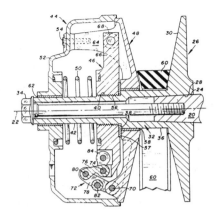 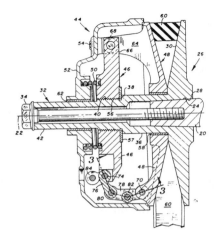

Prob 4.8.

Appendix A
Linearization of the Weibull Equation

The Weibull Equation Eq. (4.11) and [1.5]

$$\frac{d[\ln R(t)]}{dt} = -h(t) = -\frac{f(t)}{1 - F(t)} \tag{A.1}$$

Integrating

$$R(t) = \exp\left[-\int_0^t h(\tau)d\tau\right] \tag{A.2}$$

if $h(\tau)$ is Weibull from Eq. (1.2) and Eq. (1.16)

$$\frac{d}{dt}[\ln R(t)] = -\frac{\beta}{\delta}(t - \gamma)^{\beta-1} \tag{A.3}$$

Integrating from γ the lowest value of the data to $t = \tau$

$$\ln R(t) = -\frac{1}{\delta}(\tau - \gamma)^{\beta}\Big|_{\gamma}^{t}$$

$$= -\frac{1}{\delta}(t - \gamma)^{\beta} \tag{A.4}$$

This equation is a natural logarithmic form of the following in the two forms Eq. (1.16) also called $Q(t)$ the failure

$$Q(t) = 1 - \exp\left[-\frac{(t - \gamma)^{\beta}}{\delta}\right] \tag{A.5}$$

$$Q(t) = 1 - \exp\left[-\left(\frac{t - \gamma}{\theta}\right)^{\beta}\right] \tag{A.6}$$

and also note $h(t)$ is a constant

$$Q(t) = 1 - \exp[-\lambda t] \tag{A.7}$$

Further noting β ranges from about 1 to higher values most generally around 5–10 for a Gaussian distribution. In all forms by definition normalized

$$R(t) + Q(t) = 1 \tag{A.8}$$

The equation is solved for the failure $Q(t)$ using Eqs. (A.5) and (A.6) with Eq. (1.4) the two forms are

$$Q(t) = 1 - \exp\left[-\frac{(t-\gamma)^\beta}{\delta}\right] \tag{A.9}$$

$$Q(t) = 1 - \exp\left[-\left(\frac{t-\gamma}{\theta}\right)^\beta\right] \tag{A.10}$$

Rearranging, taking the natural logarithm twice, and noting $\ln e = 1$

$$\ln\ln\left[\frac{1}{1-Q(t)}\right] = \beta\ln(t-\gamma) - \ln\delta \tag{A.11}$$

$$\ln\ln\left[\frac{1}{1-Q(t)}\right] = \beta\ln\left(\frac{t-\gamma}{\theta}\right) \tag{A.12}$$

Weibull paper as used in Chapter 1 Examples may be used for a graphical representation and values of β and δ or θ obtained assuming γ is the actual lowest number. These are crude considering the SAS computer uses several runs to obtain final results. Here again and explained in Chapter 1 the value for γ is related to half of γ or even zero to match the graphical solution on Weibull paper. In fact running three runs with $\gamma^1 = 0$, $\gamma/2$, and γ would allow comparison of three separate runs to see if the βs and δs or θs change.

Appendix B
Monte Carlo Calculations

I. MONTE CARLO SIMULATIONS

The simulation procedure can be broken down into seven steps:

1. Fit failure criterion data (usually yield strength or tensile strength) to an appropriate distribution function. Goodness-of-fit statistics are useful in the determination of an acceptable model.
2. Define the applied stress on the part to be designed.
3. Assign a distribution function to each variable in the stress equation and assume a starting value for each, variables are typically load and dimensions.
4. Generate random variates from the failure criterion distribution and from each of the variable distributions.
5. Calculate the stress using the random variate for each variable and compare that stress with the random variate from the failure criterion distribution. Whenever the stress exceeds the failure variate, a failure has occurred.
6. Repeat the last two steps n times, where $1/n =$ probability of failure; e.g., a probability of failure of 10^{-6} requires 10^6 calculations.
7. If only one failure has occurred in the n calculations, the design is valid for a probability failure of $1/n$. If no failure occurred, or more than one occurred, adjust the assumed design variables (step 4) and repeat last three steps until only one failure occurs in n simulations.

II. GENERATING RANDOM VARIATES

Most computer programming languages are capable of generating a random variate from a uniform distribution where every real number on an interval $0 \leq n \leq N$ has an equal probability of being randomly chose (N is generally 1).

The random number generator uses a seed to initialize the process. If the user does not provide a value for the seed, the computer will use it's internal clock for initialization. Each call to the random number generator then creates a random number and a new seed for the next call.

Computer generated random numbers are commonly referred to as pseudo-random variates since the computer will generate the same sequence of numbers, if it is given the same seed for initialization.

Unless the user knows how the computer arrives at its internal seed, it is strongly suggested that an external seed be used. The last generated seed is then saved and used to initialize the random number generator on the next application.

III. GENERATING RANDOM VARIATES FROM OTHER DISTRIBUTIONS

Many algorithms for generating other random variates from a uniform random variate are widely available in the literature. A collection of usable algorithms can be found in Rubenstein [B1].

REFERENCES

B1. Rubenstein RY. Simulation and the Monte Carlo Method, John Wiley and Sons, 1981.

Appendix C
Computer Optimization Routines
[3.20]

Optimization problems can be categorized as:

I. UNCONSTRAINED MINIMIZATION OF THE CRITERION FUNCTION

$$\text{minimize } C(x_i) \tag{C.1}$$

where

1. The criterion function is continuous and can be either linear or non-linear
2. The function is not automatically minimized for all the variables equal to zero
3. Negative solutions must be assigned a meaning or ignored

II. CRITERION FUNCTION WITH SIMPLE REGIONAL CONSTRAINTS

$$\text{minimize } C(x_i) \tag{C.2}$$

with

$$L_i \leq x_i \leq U_i \quad \text{for} \quad i = 1, 2, 3, \ldots, n \tag{C.3}$$

where

1. The criterion function is continuous and can be either linear or non-linear
2. Each x_i is bounded by L_i and U_i. L_i or U_i could be zero. Not all x_i have to be constrained
3. The criterion function is not automatically minimized for all the variables equal to zero

III. CRITERION FUNCTION WITH LINEAR FUNCTIONAL CONSTRAINTS

$$\text{minimize } C(x_i) \quad \text{for} \quad i = 1, 2, \ldots n \tag{C.4}$$

with

$$F_k = \sum_j A_{jk} x_j + B_k = 0 \tag{C.5}$$

where

    1.   The criterion function is continuous and can be either linear or non-linear

    2.   The criterion function is not automatically minimized for all the variables equal to zero

IV. CRITERION FUNCTION WITH NON-LINEAR CONSTRAINTS

$$\text{minimize } C(x_i) \tag{C.6}$$

with

$$
\begin{aligned}
F_j(x_j) &= 0 \quad \text{for} \quad j = 1, 2, \ldots m \\
F_j(x_j) &\geq 0 \quad \text{and} \quad j = m + 1, \ldots, p
\end{aligned}
\tag{C.7}
$$

and

$$x_i = 1, 2, \ldots, n$$

where

    1.   The criterion function and function constraints are continuous

    2.   The criterion function when properly constrained should not automatically minimize when all the variables are zero

V. TECHNIQUES FOR SOLUTION

There are dozens of algorithms for optimizing functions–none will work for all cases, and all find local minima. The global minimum can be inferred by finding local minima over a realistic range of the variables. Not all functions will have a global minimum.

All of the techniques are iterative in nature and require repeated calculations of the criterion function, the gradient function, the Hessian matrix (second order partial derivatives), values of constraints, and the Jacobian matrix (first-order partial derivatives) of the constraint functions.

Not all techniques require the use of derivatives and some algorithms use approximations instead of the derivatives. If more than one algorithm can be used for an application, each should be used as a check.

A good review of algorithms is given by [C.1]

REFERENCES

C.1.   Moré JJ, Wright SJ. Optimization Software Guide. Philadelphia, PA: SIAM Society for Industrial and Applied Mathematics, 1993.

Appendix D
Mechanical Failure Rates for Non-Electronic Reliability

I. SOURCES FOR INFORMATION

The task of finding failure rates can be a difficult one. Older reports are filed in

> Defense Technical Information Center
> Cameron Station
> Alexandria, VA 223214-6145, USA

These reports have been written by and for the defense industry in order to estimate overall system reliability.

The failure rates are also filed in a computer data base at

> Government Industry Exchange Program (GIDEP)
> GIDEP Operations Center
> P.O. Box 8000
> Corona, CA 91718-8000, USA

Failure rates are stated in test reports for a specific system which can be time consuming when looking for an overall failure for a class of systems. GIDEP is better known for the notices for obtaining hard-to-find mechanical or electric components to maintain and manufacture military or government systems. Alerts are also issued for components which are not being manufactured to specifications. Here all future systems in the field must be checked to ensure proper performance. GIDEP also holds seminars twice a year for users to understand the program.

One of the most concentrated efforts is at Grifiss Air Force Base at

> The Reliability Analysis Center
> P.O. Box 4700
> Rome, NY 13442-4700, USA

The center offers publications and maintains a consulting staff and publishes a Reliability Journal [4.19] and a 1000 page report [4.18] in NPRD 95 "Non Electronic Parts Reliability Data" which gathers data from several sources. In this appendix excerpts from [4.17] Tables D.6 and D.7 are presented as well as a reliability Section II from [4.21] published with permission.

The reference [4.30] offers failure rates for the marine industry and these are in Table 4.1.

When the failure rates are selected the environment is important. The second item is the number of items and hours to develop the failure. One item is how long a system can be expected to operate. Table D.1 is a good guide from bearing life estimates. It should be noted bearing life and MTTF need not be the same. So that

$$R(t) = e^{-\lambda t} \tag{D.1}$$

set $R(t)$ to say, 0.90 for life 100,000 hr Table D.1 and solve for the λ and MTTF. The number may or may not be possible for the physical system.

The following Sections II and III are published with permission of the McGraw-Hill companies from *Mechanical Design and Systems Handbook*, Harold A. RothBart, Editor, McGraw-Hill, Inc., 1964. The chapter cited is 18.11 System Reliability Analysis by Carl H. Levinson. The source for Section II is "Generic Failure Rates," Martin Company Report, Baltimore, Md.

Table D.1 Bearing-life recommendations for various classes of machinery

Type of application	Life, kh
Instruments and apparatus for infrequent use	Up to 0.5
Aircraft engines	0.5–2
Machines for short or intermittent operations where service interruption is of minor importance	4–8
Machines for intermittent service where reliable operation is of great importance	8–14
Machines for 8-hr service which are not always fully utilized	14–20
Machines for 8-hr service which are fully utilized	20–30
Machines for continuous 24-hr service	50–60
Machines for continuous 24-hr service where reliability is of extreme importance	100–200

Reproduction with permission of the McGraw-Hill companies from J. Shigley and C.R. Mischke, Mechanical Engineering Design, 5th Edition 1989.

Section III is from Earles, D., Eddins, M., and Jackson D. A theory of component part life expectancies, 8th National Symposium on Reliability and Quality Control, 1962.

II. FAILURE-RATE TABULATION

Table D.3 is a comprehensive tabulation of many different components and their failure rates and has been compiled from an analysis of component performance in actual applications. It should be noted that the following severity factors K_F, Table D.2, must be used in applying the failure rates in order to take into account the effects of environment: Therefore, the failure rate

$$\lambda = \lambda_G K_F \tag{D.2}$$

III. COMPONENT LIFE EXPECTATION

Table D.5 gives a comprehensive tabulation of life expectancies for many different components. Systems-maintenance procedures should be based upon these life expectancies. Where maintenance is not practical, such as in space-vehicle applications, alternate-mode operation with appropriate switching must be provided. This is necessary if the design cannot be changed so as to utilize a longer-life component. It should be noted that the life expectancy severity factors K_L from Table D.4 must be used in determining component life expectancies in order to take into account the effects of the environment. Therefore, the wear-out life

$$t_w = t_G K_L \tag{D.3}$$

Table D.2 Severity factors K_F

Laboratory computer	1
Ground equipment	10
Shipboard equipment	20
Trailer-mounted equipment	25
Rail-mounted equipment	30
Aircraft equipment (bench test)	50
Missile equipment (bench test)	75
Aircraft equipment (in flight)	100
Missile equipment (in flight)	1000

Table D.3 Generic failure-rate distributions λ_G

Component or part	Upper extreme	$\lambda_G / 10^6$ hr mean	Lower extreme
Absorbers, *r-f*	1.20	0.687	0.028
Accelerometers	7.5	2.8	0.35
Accelerometers, strain gage	21.4	8.0	1.00
Accumulator	19.3	7.2	0.40
Actuators	13.7	5.1	0.35
Actuators, booster servo	33.6	12.5	0.86
Actuators, sustainer servo	33.6	12.5	0.86
Actuators, small utility	9.6	3.6	0.17
Actuators, large utility	18.5	6.9	0.60
Adapters, bore-sight	6.53	2.437	0.01
Adapters, wave-guide	9.31	3.475	0.139
Alternators	2.94	0.7	0.033
Antennas	3.52	2.0	0.48
Antenna drives	10.04	5.7	1.36
Attenuators	1.30	0.6	0.15
Base castings	0.70	0.175	0.015
Baffles	1.3	1.0	0.12
Batteries, chargeable	14.29	1.4	0.5
Batteries, one shot	300 cycles	30 cycles	10 cycles
Bearings	1.0	0.5	0.02
Bearings, ball, high-speed heavy-duty	3.53	1.8	0.072
Bearings, ball, low-speed light-duty	1.72	0.875	0.035
Bearings, rotary, sleeve-type	1.0	0.5	0.02
Bearings, rotary, roller	1.0	0.5	0.02
Bearings, translatory, sleeve shaft	0.42	0.21	0.008
Bellows	4.38	2.237	0.090
Bellows, motor in excess of 0.5 in stroke	5.482	2.8	0.113
Bellows, null-type	5.879	3.0	0.121
Blowers	3.57	2.4	0.89
Boards, terminal	1.02	0.0626	0.01
Bolts, explosive	400 cycles	40 cycles	10 cycles
Brackets, bore-sight	0.05	0.0125	0.003
Brackets, mounting	0.05	0.0125	0.003
Brackets, miscellaneous	0.55	0.1375	0.034
Bracket assemblies	7.46	2.1	0.94
Brushes, rotary devices	1.11	0.1	0.04
Bulbs, temperature	3.30	1.0	0.05
Bumpers, ring assembly	0.073	0.0375	0.002
Bumper ring supports (bracket)	2.513	1.2875	0.052
Bushings	0.08	0.05	0.02
Buzzers	1.30	0.60	0.05

Table D.3 *continued*

Component or part	Upper extreme	$\lambda_G/10^6$ hr mean	Lower extreme
Cabinet assemblies	0.330	0.03	0.003
Cable assemblies	0.170	0.02	0.002
Cams	0.004	0.002	0.001
Circuit breakers	0.04	0.1375	0.045
Circuit breakers, thermal	0.50	0.3	0.25
Clamshell, plug-in assemblies	0.70	0.175	0.10
Clutches	1.1	0.04	0.06
Clutches, magnetic	0.93	0.6	0.45
Clutches, slip	0.94	0.3	0.07
Connectors, electrical	0.47/pin	0.2/pin	0.03/pin
Connectors, AN type	0.385/pin	0.2125/pin	0.04/pin
Counters	5.25	4.2	3.5
Counterweights, large	0.545	0.3375	0.13
Counterweights, small	0.03	0.0125	0.005
Coolers	7.0	4.20	1.40
Couplers, directional	3.21	1.6375	0.065
Couplers, rotary	0.049	0.025	0.001
Couplings, flexible	1.348	0.6875	0.027
Couplings, rigid	0.049	0.025	0.001
Covers, bore-sight adapter	0.347	0.1837	0.02
Covers, dust	0.01	0.006	0.002
Covers, protective	0.061	0.038	0.015
Crankcases	1.8	0.9	0.10
Cylinders	0.81	0.007	0.005
Cylinders, hydraulic	0.12	0.008	0.005
Cylinders, pneumatic	0.013	0.004	0.002
Delay lines, fixed	0.25	0.1	0.08
Delay lines, variable	4.62	3.00	0.22
Diaphragms	9.0	6.00	0.10
Differentials	0.168	0.04	0.012
Diodes	1.47	0.2	0.16
Disconnects, quick	2.1/pin	0.4/pin	0.09/pin
Drives, belt	15.0	3.875	0.142
Drives, direct	5.26	0.4	0.33
Drives, constant-speed, pneumatic	6.2	2.8	0.3
Driving-wheel assemblies	0.1	0.025	0.02
Ducts, blower	1.3	0.5125	0.21
Ducts, magnetron	3.0	0.075	0.04
Dynamotors	5.46	2.8	1.15
Fans, exhaust	9.0	0.225	0.21
Filters, electrical	3.00	0.345	0.140

Table D.3 *continued*

Component or part	Upper extreme	$\lambda_G/10^6$ hr mean	Lower extreme
Filters, light	0.80	0.20	0.12
Filters, mechanical	0.8	0.3	0.045
Fittings, mechanical	0.71	0.1	0.04
Gaskets, cork	0.077	0.04	0.003
Gaskets, impregnated	0.225	0.1375	0.05
Gaskets, monel mesh	0.908	0.05	0.0022
Gaskets, O-ring	0.03	0.02	0.01
Gaskets, phenolic	0.07	0.05	0.01
Gaskets, rubber	0.03	0.02	0.011
Gages, pressure	7.8	4.0	0.135
Gages, strain	15.0	11.6	1.01
Gears	0.20	0.12	0.0118
Gearboxes, communications	0.36	0.20	0.11
Gears, helical	0.098	0.05	0.002
Gears, sector	1.8	0.9125	0.051
Gears, spur	4.3	2.175	0.087
Gear trains (communications)	1.79	0.9	0.093
Generators	2.41	0.9	0.04
Gimbals	12.0	2.5	1.12
Gyros	7.23	4.90	0.85
Gyros, rate	11.45	7.5	3.95
Gyros, reference	25.0	10.0	2.50
Hardware, miscellaneous	0.121	0.087	0.0035
Heaters, combustion	6.21	4.0	1.112
Heater elements	0.04	0.02	0.01
Heat exchangers	18.6	15.0	2.21
Hoses	3.22	2.0	0.05
Hoses, pressure	5.22	3.9375	0.157
Housings	2.05	1.1	0.051
Housings, cast, machined bearing surface	0.91	0.4	0.016
Housings, cast, tolerances 0.001 in or wider	0.041	0.0125	0.0005
Housings, rotary	1.211	0.7875	0.031
Insulation	0.72	0.50	0.011
Iris, wave-guide	0.08	0.0125	0.003
Jacks	0.02	0.01	0.002
Joints, hydraulic	2.01	0.03	0.012
Joints, mechanical	1.96	0.02	0.011
Joints, pneumatic	1.15	0.04	0.021
Joints, solder	0.005	0.004	0.0002
Joints, solder	0.08	0.04	0.02
Lamps	35.0	8.625	3.45

Table D.3 *continued*

Component or part	Upper extreme	$\lambda_G/10^6$ hr mean	Lower extreme
Lines and fittings	7.80	0.02	0.05
Motors	7.5	0.625	0.15
Motors, blower	5.5	0.2	0.05
Motors, electrical	0.58	0.3	0.11
Motors, hydraulic	7.15	4.3	1.45
Motors, servo	0.35	0.23	0.11
Motors, stepper	0.71	0.37	0.22
Mounts, vibration	1.60	0.875	0.20
Orifices, bleeds fixed	2.11	0.15	0.01
Orifices, variable area	3.71	0.55	0.045
Pins, grooved	0.10	0.025	0.006
Pins, guide	2.60	1.625	0.65
Pistons, hydraulic	0.35	0.2	0.08
Pumps	24.3	13.5	2.7
Pumps, engine-driven	31.3	13.5	3.33
Pumps, electric drive	27.4	13.5	2.9
Pumps, hydraulic drive	45.0	14.0	6.4
Pumps, pneumatic driven	47.0	14.7	6.9
Pumps, vacuum	16.1	9.0	1.9
Regulators	5.54	2.14	0.70
Regulators, flow and pressure	5.54	2.14	0.70
Regulators, helium	5.26	2.03	0.65
Regulators, liquid oxygen	7.78	3.00	0.96
Regulators, pneumatic	6.21	2.40	0.77
Relays, general-purpose	0.48/cs	0.25/cs	0.10/cs
Resistors, carbon deposit	0.57	0.25	0.11
Resistors, fixed	0.07	0.03	0.01
Resistors, precision tapped	0.292	0.125	0.041
Resistors, WW, accurate	0.191	0.091	0.052
Resolvers	0.07	0.04	0.02
Rheostats	0.19	0.13	0.07
Seals, rotating	1.12	0.7	0.25
Seals, sliding	0.92	0.3	0.11
Sensors, altitude	7.50	3.397	1.67
Sensors, beta-ray	21.30	14.00	6.70
Sensors, liquid-level	3.73	2.6	1.47
Sensors, optical	6.66	4.7	2.70
Sensors, pressure	6.6	3.5	1.7
Sensors, temperature	6.4	3.3	1.5
Servos	3.4	2.0	1.1
Shafts	0.62	0.35	0.15

Table D.3 *continued*

Component or part	Upper extreme	$\lambda_G/10^6$ hr mean	Lower extreme
Shields, bearing	0.14	0.0875	0.035
Shims	0.015	0.0012	0.0005
Snubbers, surge dampers	3.37	1.0	0.3
Springs	0.221	0.1125	0.004
Springs, critical to calibration	0.42	0.22	0.009
Springs, simple return force	0.022	0.012	0.001
Starters	16.1	10.0	3.03
Structural sections	1.35	1.0	0.33
Suppressors, electrical	0.95	0.3	0.10
Suppressors, parasitic	0.16	0.09	0.02
Switches	0.14/cs	0.5/cs	0.009/cs
Synchros	0.61	0.35	0.09
Synchros, resolver	1.94	1.1125	0.29
Tachometers	0.55	0.3	0.25
Tanks	0.27	0.15	0.083
Thermisters	1.40	0.6	0.20
Thermostats	0.14	0.06	0.02
Timers, electronic	1.80	1.2	0.24
Timers, electromechanical	2.57	1.5	0.79
Timers, pneumatic	6.80	3.5	1.15
Transducers	45.0	30.0	20.0
Transducers, pressure	52.2	35.0	23.2
Transducers, strain gage	20.0	12.0	7.0
Transducers, thermister	28.00	15.0	10.0
Transformers	0.02	0.2	0.07
Transistors	1.02	0.61	0.38
Tubes, electron, commercial, single-diode	2.20	0.80	0.24
Turbines	16.67	10.0	3.33
Valves	8.0	5.1	2.00
Valves, ball	7.7	4.6	1.11
Valves, blade	7.4	4.6	1.08
Valves, bleeder	8.94	5.7	2.24
Valves, butterfly	5.33	3.4	1.33
Valves, bypass	8.13	5.88	1.41
Valves, check	8.10	5.0	2.02
Valves, control	19.8	8.5	1.68
Valves, dump	19.0	10.8	1.97
Valves, four-way	7.22	4.6	1.81
Valves, priority	14.8	10.3	7.9
Valves, relief	14.1	5.7	3.27
Valves, reservoir	10.8	6.88	2.70

Table D.3 *continued*

Component or part	Upper extreme	$\lambda_G/10^6$ hr mean	Lower extreme
Valves, shutoff	10.2	6.5	1.98
Valves, selector	19.7	16.0	3.70
Valves, sequence	81.0	4.6	2.10
Valves, spool	9.76	6.9	2.89
Valves, solenoid	19.7	11.0	2.27
Valves, three-way	7.41	4.6	1.87
Valves, transfer	1.62	0.5	0.26
Valves, vent and relief	15.31	5.7	3.41

Table D.4 Life-expectancy severity factor K_L

Installation environment	All equipment	Electronic and electrical equipment	Electro-mechanical equipment	Dynamic mechanical equipment
Satellites	2.50	2.60	2.40	2.10
Laboratory computer	1.00	1.00	1.00	1.00
Bench test	0.54	0.55	0.51	0.50
Ground	0.30	0.31	0.26	0.25
Shipboard	0.19	0.21	0.17	0.15
Aircraft	0.16	0.18	0.14	0.12
Missiles	0.15	0.17	0.13	0.11

IV. CONSTANT FAILURE RATE DATA

The failure rates in Section IV are condensed from RADC non electronic reliability notebook, RADC-TR-85-194, October 1985, Ray E. Schafer et al. The notation is retained from the original document. The λ values may be corrected using Eq. (D.3) for more severe environment than listed in Table D.7. The environment is defined in Table D.6.

Table D.5 Generic life-expectancy distributions t_G

Component or part	Upper extreme	Mean $t_G/10^6$	Lower extreme
Accelerometers	0.052 hr,	0.02 hr,	0.001 hr,
	0.5 cycle	0.1 cycle	0.02 cycle
Accumulators	0.1 hr,	0.06 hr,	0.02 hr,
	0.01 cycle	0.008 cycle	0.002 cycle
Actuator, electric counter		0.1 cycle	
Actuator, linear		0.2 cycle,	
		0.0004 hr	
Actuator, rotary	10.0 cycles	0.50 cycle	0.001 cycle
Air-conditioning unit	0.0017 hr	0.004 hr,	0.007 hr
		0.15 cycle	
Alternators	0.01 hr	0.009 hr	0.001 hr
Amplifier, signal transistor, a–c		0.004 hr	
Antenna drivers	0.013 hr	0.008 hr	0.001 hr
Antennas, plasma sheath	0.001 hr	0.000050 hr	0.000040 hr
Antenna switch		0.04 hr	
Attenuators	0.01 hr	0.005 hr	0.0025 hr
		0.0036 hr	
Auxiliary power units	0.001 hr	0.000500 hr	0.0000013 hr
Batteries, chargeable	0.005 cycle	0.002 cycle	0.00003 cycle
Battery, lead-acid		0.0015 cycle	
Battery, primary type	0.016 hr	0.12 hr	0.0008 hr
Beacon, S-band radar		0.040 hr	
Bearings, dry	0.0012 hr	0.0005 hr	0.00001 hr
Bearings, lubricated	0.02 hr	0.007 hr	0.002 hr
Bearings, ball	0.016 hr	0.006 hr	0.0005 hr
Bearings, ball, midget		0.008 hr	
Bearings, ball, precision		0.04 hr	
Bearings, ball, turbine	0.001 hr	0.00065 hr	0.00005 hr
Bearings, clutch release		3.0 cycles	
Bearings, heavy-duty, lub	0.003 hr	0.002 hr	0.0001 hr
Bearings, light-duty, lub	0.01 hr	0.006 hr	0.001 hr
Bearings, precision, lub	0.05 hr	0.020 hr	0.005 hr
Bearings, rotary, roller, lub		0.052 hr	
Bearings, stagger roller		0.016 hr	
Bearings, tracker roller		0.6 cycle	
Bellows, plastic	0.050 cycle	0.01 cycle	0.001 cycle
Bellows, steel and beryllium			
copper	10.0 cycles	0.1 cycle	0.001 cycle
Bellows, aluminium and			
magnesium	0.1 cycle	0.01 cycle	0.001 cycle
Blowers	0.008 hr	0.004 hr	0.001 hr

Table D.5 *continued*

Component or part	Upper extreme	Mean $t_G/10^6$	Lower extreme
Blowers, vane, axial	0.005 hr	0.0025 hr	0.001 hr
Brake, assembly		0.020 hr	
Brushes, rotary device	0.01 hr	0.003 hr	0.001 hr
Buckets, turbine wheel		0.0065 hr	
Buzzers	1.0 cycle	0.1 cycle	0.01 cycle,
	0.0025 hr	0.001 hr	0.0005 hr
Cameras (slit)	0.002 hr	0.001 hr	0.0005 hr
Cams	100.0 cycles	1.0 cycle	0.1 cycle
Capacitor, ceramic	0.02 hr	0.016 hr	0.002 hr
Capacitor, ceramic variable		0.016 hr	
Capacitors, electrolytic	0.012 hr	0.01 hr	0.004 hr
Cells, solar	0.02 hr	0.003 hr	0.0005 hr
Choppers	0.008 hr	0.005 hr	0.0012 hr
Chopper, synchro		0.026 hr	
Chopper, microsignal		0.004 hr	
Circuit breakers	0.05 cycle	0.035 cycle	0.01 cycle
Clutch, high-speed backstopping		0.019 hr	
Clutch, precision indexing		80.0 cycles	
Coatings, vitrous ceramic	0.0006 hr	0.00045 hr	0.0001 hr
Commutators	0.014 hr	0.006 hr	0.001 hr
Compressors (lub. bearings)	0.02 hr	0.007 hr	0.002 hr
Compressor diaphragm (oil-free)		0.003 hr	
Connector, electrical	0.024 hr	0.019 hr	0.012 hr,
	0.0005 cycle	0.0001 cycle	0.00005 cycle
Contactors	0.1 cycle	0.05 cycle	0.02 cycle
Contactors, power		0.05 cycle	
Contractor, rotary power		0.1 cycle	
Control package, pneumatic		0.02 hr	
Converter, analog	0.004 hr	0.003 hr	0.002 hr
Counters to digital	0.6 cycle	0.003 cycle	0.1 cycle
Counters, electronic	100.0 cycles	60.0 cycles	30.0 cycles
Counter, Geiger		0.0034 hr	
Counter, heavy-duty		0.2 cycle	
Counter, magnetic		400.0 cycles	
Cylinder, piston 250 psi, air	50.0 cycles	30.0 cycles	12.0 cycles
Delay lines, fixed	0.025 hr	0.005 hr	0.01 hr
Delay lines, variable		0.010 hr	
Detection mechanism, star wheel		0.10 cycle	
Detectors, micrometeorite	0.05 hr	0.01 hr	0.001 hr

Table D.5 *continued*

Component or part	Upper extreme	Mean $t_G/10^6$	Lower extreme
Diaphragms, Teflon	0.01 cycle	0.005 cycle	0.001 cycle
Differentials		1.0 cycle	
Digital telemetry extraction equipment		0.007 hr	
Diodes, semiconductor		0.20 hr	
Drive assembly		0.016 hr	
Drive, belt		0.002 hr	
Drives, direct lub bearing	0.02 hr	0.01 hr	0.002 hr
Drive rotary solenoid		5.0 cycles	
Equalizer, pressure-bellows type		100.0 cycles	
Fastener, Nylatch		0.06 cycle	
Fasteners, socket head	0.45 cycles	0.4 cycle	0.3 cycle
Fastener, threaded (bolt and nut)		8.0 cycles	
Filter, electrical	0.02 hr	0.012 hr	0.008 hr
Filter, pneumatic		0.040 hr	
Gaskets, rubber nonworking	0.044 hr	0.035 hr	0.026 hr
Gaskets, phenolic	0.088 hr	0.035 hr	0.018 hr
Gages, pressure	10.0 cycles	0.1 cycle	0.001 cycle
Gages, electric field and ion		0.04 hr	
Gage, gas density		0.04 hr	
Gear boxes, communications	0.02 hr	0.005 hr	0.002 hr
Gear head		0.001 hr	
Gears, steel	10.0 cycles	1.0 cycle	0.1 cycle
Gear trains, communications	0.0025 hr	0.0018 hr	0.001 hr
Generators	0.02 hr	0.01 hr	0.002 hr
Generator, brushless synchronous motor		0.010 hr	
Generators, d–c	0.02 hr	0.006 hr	0.002 hr
Generator, phase		0.03 hr	
Generators, reference	0.02 hr	0.008 hr	0.0015 hr
Generator, solid propellant		0.00004 sec	
Gyros, rate	0.040 hr	0.001 hr	0.0001 hr
Gyros, reference	0.002 hr	0.001 hr	0.0002 hr
Heater elements	0.016 hr	0.012 hr	0.001 hr
Hoses, flex	0.8 cycle	0.5 cycle	0.25 cycle
Hoses, plastic, metal-braided	0.5 cycle	0.22 cycle	0.05 cycle

Table D.5 *continued*

Component or part	Upper extreme	Mean $t_G/10^6$	Lower extreme
Indicator, elapsed time		0.01 cycle	
Inductor		0.15 hr	
Inertial reference package		0.02 hr	
Integrating hydraulic package		0.02 hr	
Inverters, 400 cps		0.04 hr	
Jack, tip		0.0001 cycle	
Joints, mechanical	1.0 cycle	0.1 cycle	0.0025 cycle
Joint, rotary		0.016 hr	
Junction box		0.04 hr	
Lamps	0.025 hr	0.02 hr	0.000005 hr
Lines and fittings	0.02 hr	0.007 hr	0.003 hr
Meters, electrical		1.0 cycle,	
	0.0135 hr	0.009 hr	0.005 hr
Meter, relay	20.0 cycles	15.0 cycles	10.0 cycles
Monitor, speed (engine)		0.01 hr	
Monitor, cosmic-ray		0.04 hr	
Motors (lub. bearings)	0.02 hr	0.007 hr	0.002 hr
Motors, blower (lub. bearings)	0.02 hr	0.007 hr	0.002 hr
Motors, electrical (lub. bearings)	0.2 hr	0.01 hr	0.001 hr
Motor, electrical, a–c		0.03 hr	
Motors, electrical, d–c	200.0 cycles	100.0 cycles	0.4 cycle
Motors, electrical, d–c torque		200.0 cycles	
Motor, electrical, d–c subminiature reversible		0.003 hr	
Motor, electrical, d–c nonferrous rotor		0.002 hr	
Motor, hydraulic		0.01 hr	
Motors, servo	0.02 hr	0.008 hr	0.001 hr
Multiplexer		0.012 hr	
Pistons, hydraulic	0.1 cycle	0.05 cycle	0.01 cycle
Power unit, electrohydraulic		0.002 hr	
Potentiometers	0.03 hr	0.02 hr	0.00025 hr
Pump		0.005 hr	
Pumps, engine-driven	0.005 hr	0.0035 hr	0.002 hr
Pumps, electric-driven	0.001 hr	0.0005 hr	0.00015 hr
Pumps, ion	0.2 hr	0.02 hr	0.002 hr
Pump, pneumatic-driven		0.003 hr	
Pump, variable-displacement (hyd.)		0.001 hr	

Table D.5 *continued*

Component or part	Upper extreme	Mean $t_G/10^6$	Lower extreme
Pump, variable displacement (hyd.) miniature	0.0005 hr	0.00025 hr	0.00003 hr
Pumps, vane, miniature		0.002 hr	
Recorders, video-tape	0.005 hr	0.002 hr	0.001 hr
Rectifiers	0.04 hr	0.021 hr	0.003 hr
Regulator, flow and pressure		0.001 hr	
Regulator, pressure pneumatic		0.04 hr	
Regulator, gas pressure		0.0015 hr	
Regulators, voltage	0.04 hr	0.02 hr	0.005 hr
Relays, general-purpose	0.7 cycle	0.2 cycle	0.02 cycle
Relays, heavy-duty	0.2 cycle	0.145 cycle	0.02 cycle
Relays, sensitive	0.3 cycle	0.2 cycle	0.02 cycle
Resistors, carbon deposit	0.05 hr	0.015 hr	0.012 hr
Resistors, composition	0.04 hr	0.014 hr	0.011 hr
Resistors, variable, composition	0.017 hr	0.016 hr	0.014 hr
Resolver		0.002 hr	
Rheostat		0.03 hr	
Ring, seal		0.001 min	
Rotor		0.01 hr	
Seals, mechanical	0.01 hr	0.007 hr	0.003 hr
Seals, electronic		0.01 hr	
Sensor, temperature		0.01 hr	
Sensors, pressure differential, bellows type	1.0 cycle	0.35 cycle	0.01 cycle
Servo	0.065 hr	0.04 hr	0.02 hr
Socket, electron tube		0.02 hr	
Solar collector		0.0035 hr	
Solar cells		0.000720 hr	
Solenoids	100.0 cycles	60.0 cycles	20 cycles
Spark plug		0.0025 hr	
Switches	0.05 cycle	0.03 cycle	0.01 cycle
Switch, cam		0.01 cycle	
Switch, miniature		0.2 cycle	
Tachometers	0.016 hr	0.01 hr	0.005 hr
Thermostats	0.5 cycle	0.25 cycle	0.02 cycle
Timers, electronic	0.02 hr	0.01 hr	0.005 hr
Timer, pneumatic		0.0025 hr	
Timer, high-precision totalizer		2.0 cycles	
Timer, elementary		0.04 hr	

Table D.5 *continued*

Component or part	Upper extreme	Mean $t_G/10^6$	Lower extreme
Transducer		0.01 hr	
Transducer, potentiometer		0.015 hr	
Transducer, strain gage		0.02 hr	
Transducer, temperature		0.005 hr	
Transducer, low-pressure		10.0 cycles	
Transducer, power		0.02 hr	
Transducers, pressure	0.1 cycle	0.05 cycle	0.02 cycle
Transformers	0.03 hr	0.01 hr	0.004 hr
Transistors		0.2 hr	
Tube, electron, receiving		0.009 hr	
Tubes, electron, power	0.01 hr	0.007 hr	0.003 hr
Turbines (limited by bearings)	0.015 hr	0.011 hr	0.008 hr
Turbines, gas	0.1 hr	0.05 hr	0.007 hr
Valves	0.2 cycle	0.15 cycle	0.06 cycle

1. Assume continuous operation.

2. Assume a theoretical laboratory computer element or component in order to comply with general trend data for actual usage element or component. Equipment that is normally not used in laboratory computers can be visualized in such an application for purposes of arriving at a generic condition.

3. Assume no adjustments other than automatic.

4. The mean and extremes do not necessarily apply to statistical samples but encompass application and equipment-type variations, also.

5. The values should be applied as operating installation conditions only and no comparisons made at the generic level.

Table D.6 Definitions of environment abbreviations

Abbreviation	Environment
AIF	Airborne, Inhabited, Fighter
AIT	Airborne, Inhabited, Transport
ARW	Airborne, Rotary Winged
AUF	Airborne, Uninhabited, Fighter
AUT	Airborne, Uninhabited, Transport
GB	Ground, Benign
GF	Ground, Fixed
GM	Ground, Mobile
ML	Missile, Launch
NS	Naval, Sheltered
NSB	Naval, Submarine
NU	Naval, Unsheltered

Table D.7   Constant failure rates condensed from [4.17]

ENV	80% λ_L Bound	Failure rate (failures per million hours) Failure rate Estimate	80% λ_U Bound
Accelerometer	Forced Balanced	Identification Number 1	
GM	8.562	26.633	37.887
Accelerometer	Pendulum, Linear	Identification Number 2	
AUF	15.559	30.408	55.906
Accelerometer	Pendulum, Single axis	Identification Number 3	
AUF	3.747	6.065	9.590
Accumulator	Hydraulic-pneumatic	Identification Number 5	
ARW	481.314	522.513	567.516
Actuator	Electromagnetic (Linear)	Identification Number 9	
GF	2.007	8.893	26.933
Actuator	/Mechanical	Identification Number 11	
AUT	1.140	5.110	15.303
Air conditioner	/Comfort	Identification Number 13	
GF	635.451	711.111	796.307
Air conditioner	/General	Identification Number 14	
GM	0.0	0.000	847.136
Air conditioner	/Process	Identification Number 15	
GF	0.0	0.000	12.876
Antenna	/Communication	Identification Number 16	
GM	2.744	6.658	14.246
Antenna (airborne)	/Microwave (communication)	Identification Number 17	
ARW	16.088	19.120	22.734
Antenna	/Radar	Identification Number 18	
GF	0.446	2.000	5.989
Axle	/General	Identification Number 19	
GM	3.932	9.539	20.410
Azimuth encoder	/Optical	Identification Number 20	
NU	9.105	40.800	122.185
Bearing	/Sleeve	Identification Number 24	
GM	3.152	4.661	6.814
Bearing nut	/General	Identification Number 25	
ARW	425.572	546.468	700.711
Belt	/Geared	Identification Number 27	
GF	0.0	0.000	117.927
Belt	/Timing	Identification Number 28	
GM	5.110	9.987	18.362
Belt	/V-Belt	Identification Number 29	

Table D.7 *continued*

ENV	80% λ_L Bound	Failure rate (failures per million hours)		
		Failure rate Estimate	80% λ_U Bound	
GM Binocular	3.752 /Nitrogen pressurized	16.812 Identification Number 30	50.348	
GM Blade assembly	607.566 /General	1058.201 Identification Number 31	1778.231	
ARW Blowers & fans	294.660 /Axial	364.312 Identification Number 32	450.364	
NS Blowers & fans	1.319 /Centrifugal	1.584 Identification Number 33	1.904	
GM Boot (dust & moisture)	2.990 /General	4.840 Identification Number 34	7.653	
ARW Brake	234.177 /Electromechanical	327.881 Identification Number 35	456.106	
GF Brushes	6.595 /Electric motor	16.000 Identification Number 36	34.234	
GF Burner	0.446 /Catalytic	2.000 Identification Number 37	5.989	
NS Bushings	471.959 /General	530.241 Identification Number 38	596.124	
GM CAM	0.654 /General	0.777 Identification Number 39	0.924	
AUT Camera	0.912 /Motion (TV)	4.088 Identification Number 40	12.242	
GF Cesium beam tube	83.686 /General	135.457 Identification Number 41	214.193	
GF Compressor	22.300 /General	34.277 Identification Number 46	51.848	
GF Compressor	1.785 /High pressure	8.000 Identification Number 47	23.958	
Compressor	/Low pressure	Identification Number 48		
NS Computer mass memory	149.846 /Fixed head disk	202.076 Identification Number 49	271.479	
NS Computer mass memory	17.853 /Magnetic tape	80.000 Identification Number 50	239.580	
NS Computer mass memory	70.938 /Moveable head disk	101.781 Identification Number 51	144.790	

Table D.7 *continued*

ENV	80% λ_L Bound	Failure rate (failures per million hours) Failure rate Estimate	80% λ_U Bound
GF	73.911	105.904	150.655
Control tube assembly	/General	Identification Number 52	
ARW	55.922	109.294	200.936
Cord/cable	/General	Identification Number 53	
NU	0.272	0.440	0.696
Counter	/Analog	Identification Number 54	
GF	3.070	6.000	11.031
Counter	/Digital	Identification Number 55	
AUT	0.0	0.000	32.899
Counter	/Mechanical	Identification Number 56	
AUF	0.0	0.000	8.157
Counter	/Water clock	Identification Number 57	
GF	34.322	55.556	87.848
Coupling	/Fluid	Identification Number 59	
NSB	64.321	85.470	113.232
Coupling	/General	Identification Number 60	
NS	0.538	1.305	2.793
Crankshaft	/General	Identification Number 62	
AUT	4.212	10.220	21.867
Cross head	/General	Identification Number 63	
ARW	30.031	72.862	155.899
Diffuser	/General	Identification Number 65	
AUT	0.0	0.000	10.966
Disc assembly	/General	Identification Number 66	
ARW	340.987	386.173	437.651
Distillation unit	/From distilling plant	Identification Number 67	
NS	345.030	483.092	672.016
Drive	/Gear	Identification Number 68	
AUT	1.140	5.110	15.303
Drive	/General	Identification Number 69	
ARW	203.130	291.449	414.605
Drive	/Variable pitch	Identification Number 70	
GF	4.836	7.151	10.454
Drive for computers tapes & discs	/Capstan motor	Identification Number 71	
GF	3.728	7.286	13.396
Drive for computer tapes & discs	/Discs	Identification Number 72	

Table D.7 *continued*

ENV	80% λ_L Bound	Failure rate (failures per million hours) Failure rate Estimate	80% λ_U Bound
GB	7.319	17.758	37.995
Drive for computer tapes & discs	/Magnetic tape transport	Identification Number 73	
NS	19.529	38.168	70.172
Drive for computer tapes & discs	/Reel motor	Identification Number 74	
GF	4.740	7.286	11.021
Drive rod	/General	Identification Number 75	
GM	0.0	0.000	27.060
Drum	/General	Identification Number 76	
GF	0.0	0.000	12.764
Duct	/General	Identification Number 77	
GF	2.023	2.902	4.128
Electric heaters	/Resistance	Identification Number 78	
NU	11.211	27.200	58.198
Electromechanical timers	/General	Identification Number 79	
AUF	2.071	4.048	7.443
Engines	/General	Identification Number 80	
ARW	1328.109	1392.106	1459.500
Feedhorn	/Waveguide	Identification Number 81	
AUT	41.485	81.000	62.695
Filter	/Gas (air)	Identification Number 82	
GM	2.192	3.242	4.739
Filter	/Liquid	Identification Number 83	
GF	3.070	6.000	11.031
Filter	/Optical	Identification Number 84	
AUT	0.912	4.088	12.242
Fittings	/General	Identification Number 85	
NS	9.105	40.800	122.185
Fittings	/Permanent	Identification Number 86	
GF	1.023	2.000	3.677
Fittings	/Quick disconnect	Identification Number 87	
GF	0.550	1.333	2.853
Flash lamp	/General	Identification Number 89	
AUT	4.561	20.440	61.212
Fuse holder	/Block	Identification Number 90	
GM	0.889	3.985	11.935
Fuse holder	/Extractor post	Identification Number 91	

Table D.7 *continued*

		Failure rate (failures per million hours)	
	80%	Failure	80%
	λ_L	rate	λ_U
ENV	Bound	Estimate	Bound
GF	1.649	4.000	8.559
Fuse holder	/Plug	Identification Number 82	
GM	0.370	1.680	4.971
Gas dryer desicator	/Molecular sieve	Identification Number 93	
NS	17.035	76.336	228.607
Gaskets and seals	/General	Identification Number 94	
NU	1.308	3.173	6.790
Gaskets & seals	/Static	Identification Number 95	
NU	0.801	1.395	2.344
Gear	/Antirotation	Identification Number 96	
GM	807.898	1578.948	2902.393
Gear	/Bevel	Identification Number 97	
GF	0.550	1.333	2.853
Gear	/Hypoid	Identification Number 99	
GF	1.116	5.000	14.974
Gear	/Worm	Identification Number 101	
AUT	0.000	0.000	8.048
Gear box	/Multiplier	Identification Number 102	
ARW	2071.857	2159.230	2250.676
GF	0.0	0.000	16.096
Gear box	/Reduction	Identification Number 103	
GF	1.116	5.000	14.974
Gear train	/Bevel	Identification Number 104	
AUF	5.126	5.813	6.596
Generator	/AC	Identification Number 105	
GM	3.932	9.539	20.410
Generator	/General (oxygen generator)	Identification Number 106	
NS	1759.922	1900.320	2052.836
Glass (sight gauge)	/General	Identification Number 107	
ARW	328.998	437.174	579.174
Grommet	/General	Identification Number 108	
AUT	0.760	3.407	10.202
Gimbals	/General	Identification Number 109	
AUT	8.425	20.440	43.734
NU	4.553	20.400	61.092
Gimbals	/Torque	Identification Number 110	
AUF	3.599	5.164	7.347
Gyroscope	/Single axis	Identification Number 111	

Table D.7 *continued*

ENV	80% λ_L Bound	Failure rate Estimate	80% λ_U Bound
		Failure rate (failures per million hours)	
AUT	391.885	449.677	516.352
Gyroscope	/Two axis rotor	Identification Number 112	
AUF	42.919	49.422	56.949
Heat exchangers	/Coplates	Identification Number 113	
GM	5.641	11.025	20.269
Heat exchangers	/General	Identification Number 114	
GF	1.319	3.200	6.847
Heat exchangers	/Radiator	Identification Number 115	
GM	2.310	5.604	11.991
Heater	/Water	Identification Number 116	
GF	363.796	422.222	490.362
Heater blankets	/General	Identification Number 117	
AUT	2.281	10.220	30.606
Heater flex element	/Heater tape	Identification Number 118	
AUF	0.423	0.736	1.237
High speed printer	/Electrostatic	Identification Number 119	
GM	630.498	705.568	790.100
High speed printer	/Impact	Identification Number 120	
GB	1.981	8.879	26.590
High speed printer	/Thermal	Identification Number 121	
GF	22.102	29.370	38.909
Hose	/Flexible	Identification Number 122	
NSB	39.185	53.763	73.397
Housing	/General	Identification Number 125	
ARW	229.535	279.309	339.940
Incinerator	/From sewage treatment	Identification Number 126	
NS	838.583	2034.588	4353.273
Instruments	/Ammeter	Identification Number 127	
GF	1.785	8.000	23.958
Instruments	/Flow meter	Identification Number 128	
GF	8.079	15.789	29.029
Instruments	/Humidity indicator	Identification Number 129	
AUT	4.561	20.440	61.212
Instruments	/Indicator	Identification Number 130	
GF	2.108	4.120	7.575
Instruments	/Indicator (light)	Identification Number 131	
NS	0.538	1.305	2.793

Table D.7 *continued*

ENV	80% λ_L Bound	Failure rate (failures per million hours) Failure rate Estimate	80% λ_U Bound
Instruments	/ Indicator (fluid level)	Identification Number 132	
GM	406.086	793.651	1459.126
Instruments	/ Pressure gauge	Identification Number 133	
AUT	26.596	40.880	61.836
Instruments	/ Time meter	Identification Number 134	
NS	1.499	2.611	4.387
Instruments	/ Total time meter	Identification Number 135	
GF	14.955	17.132	19.640
Instruments	/ Voltmeter	Identification Number 136	
NS	1.053	1.704	2.694
Joint microwave rotary	/ General	Identification Number 137	
GM	4.867	7.878	12.457
Keyboard	/ Electromechanical	Identification Number 138	
GB	1.981	8.879	26.590
Keyboard	/ General	Identification Number 139	
GF	4.495	6.909	10.451
Keyboard	/ Mechanical	Identification Number 140	
GF	2.679	4.866	7.841
Knob	/ General	Identification Number 141	
NS	0.298	0.722	1.545
Lamp	/ Xenon	Identification Number 142	
NS	6503.187	7704.156	9131.840
Lamp holder	/ General	Identification Number 143	
NS	0.910	1.584	2.663
Lens	/ Optical	Identification Number 144	
AUT	0.0	0.000	2.742
Low speed printer	/ Dot matrix	Identification Number 145	
GF	244.654	325.097	430.693
Manifold	/ General	Identification Number 146	
NS	0.538	1.305	2.793
GF	0.550	1.333	2.853
Metal tubing	/ General	Identification Number 147	
GF	0.077	0.150	0.278
Modules	/ General	Identification Number 148	
ARW	521.754	557.491	595.895
Motor generator set	/ AC	Identification Number 149	
GM	0.0	0.000	425.209
NS	17.035	76.336	228.607

Table D.7 *continued*

ENV	80% λ_L Bound	Failure rate (failures per million hours)		
		Failure rate Estimate	80% λ_U Bound	
Motor generator set NS	/DC 8.518	Identification Number 150 38.168	114.303	
Motor generator set AUF	/General 22.261	Identification Number 151 25.384	28.964	
Motor, Electric GF	/>1 Horse power, AC 0.550	Identification Number 152 1.333	2.853	
Motor, Electric GF	/>10 Horse power, AC 2.480	Identification Number 153 11.111	33.275	
Motor Electric NS	/DC 14.110	Identification Number 154 18.510	24.225	
Motor, Electric GB	/DC (4 horsepower) 0.0	Identification Number 155 0.000	83.323	
Motor, Electric GF	/Hydraulic, DC 2.480	Identification Number 156 11.111	33.275	
Motor, Electric GF	/Servo, DC 7.183	Identification Number 157 10.058	13.991	
Motor, Electric AUF	/Stepper 6.219	Identification Number 158 7.986	10.240	
O-Ring AUT	/General 15.823	Identification Number 159 17.629	19.354	
Particle separator ARW	/General 858.023	Identification Number 160 921.859	990.418	
Pitch horn ARW	/General 383.404	Identification Number 161 437.178	498.245	
Plotter NS	/Electromechanical 4.816	Identification Number 162 6.909	9.829	
Power circuit breaker GF	/Current & voltage trip 7.363	Identification Number 163 9.930	13.340	
Power circuit breaker NS	/Current trip 1.819	Identification Number 164 2.796	4.230	
Power switch gear GM	/General 0.734	Identification Number 165 3.288	9.846	
Precipitator NS	/Electrostatic 184.230	Identification Number 166 360.057	661.965	
Prism AUT	/Optical 0.0	Identification Number 167 0.000	10.966	

Table D.7 *continued*

	80% λ_L	Failure rate (failures per million hours) Failure rate	80% λ_U
ENV	Bound	Estimate	Bound
Propeller	/General (from ship)	Identification Number 168	
MS	368.167	893.256	1911.242
Proportioning unit	/From distilling plant	Identification Number 169	
NS	0.0	0.000	102.357
Pulley	/Gear belt	Identification Number 170	
GF	3.439	5.287	7.997
Pulley	/Grooved	Identification Number 171	
NS	0.538	1.305	2.793
Pulley	/V-Pulley	Identification Number 172	
GM	6.452	12.609	23.182
Pump	/Centrifugal	Identification Number 173	
NS	32.693	37.279	42.537
Pump	/Rotary	Identification Number 177	
NS	60.857	272.702	816.675
Pump	/Vacuum	Identification Number 178	
GF	5.429	10.610	19.507
Pump	/Vacuum-lobe type	Identification Number 179	
GF	199.298	244.444	298.946
Pump	/Vacuum-ring seal type	Identification Number 180	
GF	2.480	11.111	33.275
Purifier	/Centrifugal	Identification Number 181	
NS	1033.057	1527.717	2233.328
Quill assembly	/General	Identification Number 182	
ARW	1186.734	1296.933	1418.125
Radome	/Microwave, antenna	Identification Number 183	
AIF	3.005	7.291	15.600
Refrigeration plant	/From air conditioning plant	Identification Number 184	
NS	95.233	107.106	120.541
Regulator	/Electrical	Identification Number 185	
GF	1.093	2.652	5.673
Regulator	/Pneumatic (vacuum breaker)	Identification Number 187	
GF	80.982	111.111	151.687
Regulator	/Pressure	Identification Number 188	

Table D.7 *continued*

		Failure rate (failures per million hours)	
	80%	Failure	80%
	λ_L	rate	λ_U
ENV	Bound	Estimate	Bound
GF	1.023	2.000	3.677
Regulator	/Temperature	Identification Number 189	
NS	17.035	76.336	228.607
Resilient mount	/General	Identification Number 190	
NS	0.744	1.295	2.176
Resilient mount	/Shock mounts	Identification Number 191	
GM	91.369	178.571	328.303
Retaining ring	/General	Identification Number 192	
NS	0.347	0.678	1.247
Seal	/General	Identification Number 193	
ARW	393.223	510.037	660.340
Seal	/Solder	Identification Number 194	
AUF	1.583	2.172	2.965
Sensors	/Water level	Identification Number 195	
GF	52.594	77.778	113.701
Sensors/transducers/transmitter/acoustic (hydrophones)		Identification Number 196	
NSB	0.0	0.000	184.244
Sensors/transduces/transmitter/airflow		Identification Number 197	
GM	216.928	526.316	1126.124
Sensors/transducers/transmitter/flow (liquid)		Identification Number 198	
AUF	4.178	10.136	21.688
Sensors/transducers/transmitter/humidity		Identification Number 199	
AUT	4.561	20.440	61.212
Sensors/transducers/transmitter/infrared		Identification Number 200	
AUT	574.805	643.855	721.631
Sensors/transducer/transmitter/motion		Identification Number 201	
GM	71.867	93.217	120.687
NS	0.0	0.000	65.669
Sensors/transducer/transmitter/ temperature		Identification Number 203	
GM	9.989	13.273	17.584
Shaft	/General	Identification Number 204	
AUT	2.761	4.809	8.082
Shock absorbers	/Combination	Identification Number 205	
AUT	5.616	13.627	29.156
Shock absorbers	/Resilient	Identification Number 206	
WU	4.175	8.160	15.002

Table D.7 *continued*

ENV	80% λ_L Bound	Failure rate (failures per million hours) Failure rate Estimate	80% λ_U Bound
Slip ring-brush	/Power and signal	Identification Number 207	
AUF	0.421	0.568	0.763
Slip rings	/General	Identification Number 208	
GF	0.149	0.667	1.996
Solenoids	/General	Identification Number 209	
NS	9.105	40.800	122.185
Solenoids	/Linear	Identification Number 210	
GF	2.480	11.111	33.275
Solenoids	/Rotary	Identification Number 211	
GF	20.947	33.906	53.614
Spring	/Compression	Identification Number 212	
GF	6.254	10.892	18.303
Spring	/General	Identification Number 213	
AIF	11.192	21.873	40.214
Spring	/Torrision	Identification Number 214	
GF	7.315	14.296	26.282
Sprocket	/General	Identification Number 215	
AUT	0.912	4.088	12.242
GF	3.517	5.693	9.002
Steamboiler	/General (from ship)	Identification Number 216	
NS	378.790	510.820	686.262
Stow pin	/General	Identification Number 217	
NU	3.131	6.120	11.252
Switch	/Coaxial (electromechanical)	Identification Number 218	
NS	26.236	35.997	49.143
Switch	/Flow (liquid)	Identification Number 219	
NS	3.415	5.050	7.313
Switch	/Interlock	Identification Number 220	
GM	410.902	631.579	955.348
Switch	/Pressure (air flow)	Identification Number 221	
GF	2.983	4.176	5.809
Switch	/Rocker	Identification Number 222	
GF	8.452	10.519	13.082
Switch	/Thermostatic	Identification Number 223	
NS	2.383	2.986	3.739
Switch	/Thumbwheel	Identification Number 224	
GF	2.792	4.006	5.899

Table D.7 *continued*

ENV	80% λ_L Bound	Failure rate (failures per million hours) Failure rate Estimate	80% λ_U Bound
Switch	/Wave guide	Identification Number 225	
GF	1.649	4.000	8.559
Switchboard control	/From oxygen generator	Identification Number 226	
NS	542.640	598.377	661.336
Syncro	/Transmitter	Identification Number 227	
NS	3.502	5.178	7.570
Syncro assembly	/General	Identification Number 228	
NS	36.723	45.701	56.838
Switch	/Rocker	Identification Number 229	
AUF	0.532	1.291	2.782
Switch	/Thermostatic	Identification Number 230	
AUT	43.795	61.320	85.300
Switch	/Thumbwheel	Identification Number 231	
NU	29.294	57.252	105.258
Switch	/Wave guide	Identification Number 232	
GF	1.785	8.000	23.958
Switchboard control	/From oxygen generator	Identification Number 233	
GM	20.568	33.292	52.643
Syncro	/Transmitter	Identification Number 234	
AUT	0.0	0.000	32.899
Syncro assembly	/General	Identification Number 235	
NS	0.812	0.988	1.203
Valve	/Pneumatic	Identification Number 243	
GF	1.649	4.000	8.559
Valve	/Solenoid operated	Identification Number 244	
GF	2.480	11.111	33.275
Valve (fill & drain)	/Hand operated plug valve	Identification Number 247	
ML	0.0	0.000	922.383
Valve (bipropellant-low thurst)/torque motor operated		Identification Number 249	
ML	0.0	0.000	5152.230
Washer	/Flat	Identification Number 250	
GM	0.152	0.165	0.180
NS	0.343	0.426	0.530
Washer	/Lock	Identification Number 251	
GM	0.097	0.116	0.138

Table D.7 *continued*

| ENV | 80% λ_L Bound | Failure rate (failures per million hours) | |
		Failure rate Estimate	80% λ_U Bound
NS	0.711	0.861	1.043
Washer	/Sherr	Identification Number 252	
NS	0.350	1.569	4.699
Washer	/Spring	Identification Number 253	
GF	0.669	1.623	3.473
Washer	/Star	Identification Number 254	
GF	0.010	0.018	0.030
Water demineralizer	/Mix-resin	Identification Number 255	
NS	17.035	76.336	228.607

Appendix E
Statistical Tables

I. ONE SIDED α_i VALUES FOR A NORMAL DISTRIBUTION

One sided α_i values, such as class 'A' and 'B' structural materials stress limits are often used for designing. In Fig. E.1 the K_i values for $n-1$ degrees of freedom are plotted so that the K_i value may be estimated. The values are substituted into the following equation

$$\alpha_i = \bar{X} - K_i S \quad i = A, B, C \tag{E.1}$$

where

$\bar{X}$ is the Gaussian mean of the data set with n values.
S is the standard deviation of the same data set.
K_A values will exceed α 99% of the time with 95% confidence.
K_B values will exceed α 90% of the time with 95% confidence.
K_C values will exceed α 99.999% of the time with 95% confidence.

When more accurate values are not available from Fig. E.1 the following equations may be used to compute K factors in lieu of using table values: [1.18]

$$K_A = 2.236 + \exp[1.34 - 0.522 \ln(n) + 3.87/n] \tag{E.2}$$

$$K_B = 1.282 + \exp[0.958 - 0.520 \ln(n) + 3.9/n]. \tag{E.3}$$

These approximations are accurate to within 0.2% of the table values for n greater than or equal to 16.

Values for n from 2 to 5 are shown in Table E.1.

II. STUDENTS t DISTRIBUTION AND χ^2 DISTRIBUTIONS

The following Tables E.2 and E.3 have been reprinted with the permission of Addison Wesley Longman Ltd.

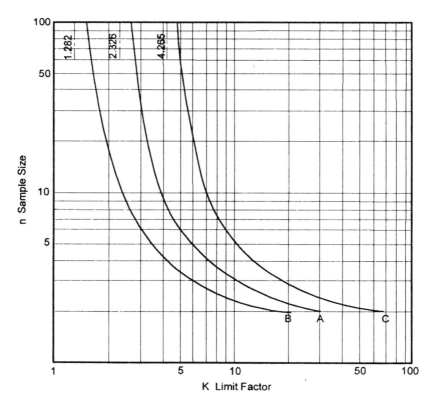

Figure E.1. One sided limit factors K_i with 95% confidence for the normal distribution and $n-1$ degrees of freedom (data plotted from ref. [1.18])

Table E.1 K_i values for n 2 to 5(1.18)

n	K_A	K_B	K_C
2	37.094	20.581	68.010
3	10.553	6.155	18.986
4	7.042	4.162	12.593
5	5.741	3.407	10.243

Table E.2 Proportions of area for the t distributions

Proportions of Area for the t Distributions

Areas reported below:

$$t = \frac{\bar{X} - \mu}{s_x}$$

Proportion of area (one tail)

df	0.10	0.05	0.025	0.01	0.005	df	0.10	0.05	0.025	0.01	0.005
1	3.078	6.314	12.706	31.821	63.657	18	1.330	1.734	2.101	2.552	2.878
2	1.886	2.920	4.303	6.965	9.925	19	1.328	1.729	2.093	2.539	2.861
3	1.638	2.353	3.182	4.541	5.841	20	1.325	1.725	2.086	2.528	2.845
4	1.533	2.132	2.776	3.747	4.604	21	1.323	1.721	2.080	2.518	2.831
5	1.476	2.015	2.571	3.365	4.032	22	1.321	1.717	2.074	2.508	2.819
6	1.440	1.943	2.447	3.143	3.707	23	1.319	1.714	2.069	2.500	2.807
7	1.415	1.895	2.365	2.998	3.499	24	1.318	1.711	2.064	2.492	2.797
8	1.397	1.860	2.306	2.896	3.355	25	1.316	1.708	2.060	2.485	2.787
9	1.383	1.833	2.262	2.821	3.250	26	1.315	1.706	2.056	2.479	2.779
10	1.372	1.812	2.228	2.764	3.169	27	1.314	1.703	2.052	2.473	2.771
11	1.363	1.796	2.201	2.718	3.106	28	1.313	1.701	2.048	2.467	2.763
12	1.356	1.782	2.179	2.681	3.055	29	1.311	1.699	2.045	2.462	2.756
13	1.350	1.771	2.160	2.650	3.012	30	1.310	1.697	2.042	2.457	2.750
14	1.345	1.761	2.145	2.624	2.977	40	1.303	1.684	2.021	2.423	2.704
15	1.341	1.753	2.131	2.602	2.947	60	1.296	1.671	2.000	2.390	2.660
16	1.337	1.746	2.120	2.583	2.921	120	1.289	1.658	1.980	2.358	2.617
17	1.333	1.740	2.110	2.567	2.898	∞	1.282	1.645	1.960	2.326	2.576

* Example. For the shaded area to represent 0.05 of the total area of 1.0, value of t with 10 degrees of freedom is 1.812.

Source: From Table III of Fisher and Yates, *Statistical Tables for Biological Agricultural and Medical Research*, 6th ed., 1974, published by Longman Group Ltd., London (previously published by Oliver and Boyd, Edinburgh), by permission of the authors and publishers.

Table E.3 Proportions of area for the χ^2 distributions

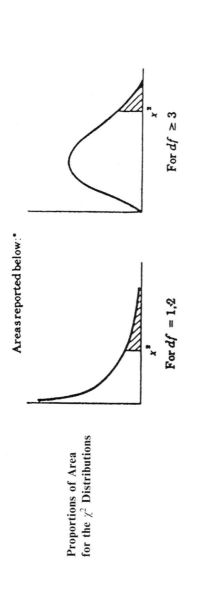

Areas reported below:

For $df = 1,2$

For $df \geq 3$

Proportions of Area for the χ^2 Distributions

df	\multicolumn{14}{c}{Proportion of area}
	0.995	0.990	0.975	0.950	0.900	0.500	0.100	0.050	0.025	0.010	0.005
1	0.00004	0.00016	0.00098	0.00393	0.0158	0.455	2.71	3.84	5.02	6.63	7.88
2	0.0100	0.0201	0.0506	0.103	0.211	1.386	4.61	5.99	7.38	9.21	10.60
3	0.072	0.115	0.216	0.352	0.584	2.366	6.25	7.81	9.35	11.34	12.84
4	0.207	0.297	0.484	0.711	1.064	3.357	7.78	9.49	11.14	13.28	14.86
5	0.412	0.554	0.831	1.145	1.61	4.251	9.24	11.07	12.83	15.09	16.75
6	0.676	0.872	1.24	1.64	2.20	5.35	10.64	12.59	14.45	16.81	18.55
7	0.989	1.24	1.69	2.17	2.83	6.35	12.02	14.07	16.01	18.48	20.28
8	1.34	1.65	2.18	2.73	3.49	7.34	13.36	15.51	17.53	20.09	21.96

Table E.3 *continued*

df	0.995	0.990	0.975	0.950	0.900	0.500	0.100	0.050	0.025	0.010	0.005
						Proportion of area					
9	1.73	2.09	2.70	3.33	4.17	8.34	14.68	16.92	19.02	21.67	23.59
10	2.16	2.56	3.25	3.94	4.87	9.34	15.99	18.31	20.48	23.21	25.19
11	2.60	3.05	3.82	4.57	5.58	10.34	17.28	19.68	21.92	24.73	26.76
12	3.07	3.57	4.40	5.23	6.30	11.34	18.55	21.03	23.34	26.22	28.30
13	3.57	4.11	5.01	5.89	7.04	12.34	19.81	22.36	24.74	27.69	29.82
14	4.07	4.66	5.63	6.57	7.79	13.34	21.06	23.68	26.12	29.14	31.32
15	4.60	5.23	6.26	7.26	8.55	14.34	22.31	25.00	27.49	30.58	32.80
16	5.14	5.81	6.91	7.96	9.31	15.34	23.54	26.30	28.85	32.00	34.27
17	5.70	6.41	7.56	8.67	10.09	16.34	24.77	27.59	30.19	33.41	35.72
18	6.26	7.01	8.23	9.39	10.86	17.34	25.99	28.87	31.53	34.81	37.16
19	6.84	7.63	8.91	10.12	11.65	18.34	27.20	30.14	32.85	36.19	38.58
20	7.43	8.26	9.59	10.85	12.44	19.34	28.41	31.41	34.17	37.57	40.00
21	8.03	8.90	10.28	11.59	13.24	20.34	29.62	32.67	35.48	38.93	41.40
22	8.64	9.54	10.98	12.34	14.04	21.34	30.81	33.92	36.78	40.29	42.80
23	9.26	10.20	11.69	13.09	14.85	22.34	32.01	35.17	38.08	41.64	44.18
24	9.89	10.86	12.40	13.85	15.66	23.34	33.20	36.42	39.36	42.98	45.56
25	10.52	11.52	13.12	14.61	16.47	24.34	34.38	37.65	40.65	44.31	46.93
26	11.16	12.20	13.84	15.38	17.29	25.34	35.56	38.89	41.92	45.64	48.29
27	11.81	12.83	14.57	16.15	18.11	26.34	36.74	40.11	43.19	46.96	49.64
28	12.46	13.56	15.31	16.93	18.94	27.34	37.92	41.34	44.46	48.28	50.99
29	13.12	14.26	16.05	17.71	19.77	28.34	39.09	42.56	45.72	49.59	52.34
30	13.79	14.95	16.79	18.49	20.60	29.34	40.26	43.77	46.98	50.89	53.67
40	20.71	22.16	24.43	26.51	29.05	39.34	51.81	55.76	59.34	63.69	66.77
50	27.99	29.71	32.36	34.76	37.69	49.33	63.17	67.50	71.42	76.15	79.49
60	35.53	37.43	40.48	43.19	46.46	59.33	74.40	79.08	83.30	88.38	91.95

Table E.3 *continued*

df	0.995	0.990	0.975	0.950	0.900	Proportion of area 0.500	0.100	0.050	0.025	0.010	0.005
70	43.28	45.44	48.76	51.74	55.33	69.33	85.53	90.53	95.02	100.4	104.2
80	51.17	53.54	51.17	60.39	64.28	79.33	98.58	101.9	106.6	112.3	116.3
90	59.20	61.75	65.65	69.13	73.29	89.33	107.6	113.1	118.1	124.1	128.3
100	67.33	70.06	74.22	77.93	82.36	99.33	118.5	124.3	129.6	135.8	140.2

* Example. For the shaded area to represent 0.05 of the total area of 1.0 under the density function, the value of χ^2 is 18.31 when $df=10$.
Source: From Table IV of Fisher and Yates, *Statistical Tables for Biological, Agricultural and Medical Research*, 6th ed., 1974, published by Longman Group Ltd., London (previously published by Oliver & Boyd, Edinburgh) by permission of the authors and publishers.

III. ORDER-STATISTIC ESTIMATES IN SMALL SAMPLES

Table E.4 Unbiased estimate of σ using the range w (variance to be multiplied by σ^2)

N	$K_1 w$	*Variance*	*Eff.*	N	$K_1 w$	*Variance*	*Eff.*
2	0.886w	0.571	1.000	11	0.315w	0.0616	0.831
3	0.591w	0.275	0.992	12	0.307w	0.0571	0.814
4	0.486w	0.183	0.975	13	0.300w	0.0533	0.797
5	0.430w	0.138	0.955	14	0.294w	0.0502	0.781
6	0.395w	0.112	0.933	15	0.288w	0.0474	0.766
7	0.370w	0.0949	0.911	16	0.283w	0.0451	0.751
8	0.351w	0.0829	0.890	17	0.279w	0.0430	0.738
9	0.337w	0.0740	0.869	18	0.275w	0.0412	0.725
10	0.325w	0.0671	0.850	19	0.271w	0.0395	0.712
				20	0.268w	0.0381	0.700

Adapted by permission of the Biometrika Trustees from E.S. Pearson and H.O. Hartley, *The Probability Integral of the Range in Samples of* n *Observations From a Normal Population*, Biometrika, Vol. 32, 1942, p. 301.

Table E.5 Modified linear estimate of σ (variance to be multiplied by σ^2)

N	*Estimate*	*Variance*	*Eff.*
2	$0.8862(X_2 - X_1)$	0.571	1.000
3	$0.5908(X_3 - X_1)$	0.275	0.992
4	$0.4857(X_4 - X_1)$	0.183	0.975
5	$0.4299(X_5 - X_1)$	0.138	0.955
6	$0.2619(X_6 + X_5 - X_2 - X_1)$	0.109	0.957
7	$0.2370(X_7 + X_6 - X_2 - X_1)$	0.0895	0.967
8	$0.2197(X_8 + X_7 - X_2 - X_1)$	0.0761	0.970
9	$0.2068(X_9 + X_8 - X_2 - X_1)$	0.0664	0.968
10	$0.1968(X_{10} + X_9 - X_2 - X_1)$	0.0591	0.964
11	$0.1608(X_{11} + X_{10} + X_8 - X_4 - X_2 - X_1)$	0.0529	0.967
12	$0.1524(X_{12} + X_{11} + X_9 - X_4 - X_2 - X_1)$	0.0478	0.972
13	$0.1456(X_{13} + X_{12} + X_{10} - X_4 - X_2 - X_1)$	0.0436	0.975
14	$0.1399(X_{14} + X_{13} + X_{11} - X_4 - X_2 - X_1)$	0.0401	0.977
15	$0.1352(X_{15} + X_{14} + X_{12} - X_4 - X_2 - X_1)$	0.0372	0.977
16	$0.1311(X_{16} + X_{15} + X_{13} - X_4 - X_2 - X_1)$	0.0347	0.975
17	$0.1050(X_{17} + X_{16} + X_{15} + X_{13} - X_5 - X_3 - X_2 - X_1)$	0.0325	0.978
18	$0.1020(X_{18} + X_{17} + X_{16} + X_{14} - X_5 - X_3 - X_2 - X_1)$	0.0305	0.978
19	$0.09939(X_{19} + X_{18} + X_{17} + X_{15} - X_5 - X_3 - X_2 - X_1)$	0.0288	0.979
20	$0.09706(X_{20} + X_{19} + X_{18} + X_{16} - X_5 - X_3 - X_2 - X_1)$	0.0272	0.978

Table E.6 Several estimates of the mean (variance to be multiplied by σ^2)

N	Median Var.	Median Eff.	Midrange Var.	Midrange Eff.	Statistic	Mean of best two Var.	Mean of best two Eff.	$(X_2+X_3+\dots+X_{N-1})/(N-2)$ Var.	$(X_2+X_3+\dots+X_{N-1})/(N-2)$ Eff.
2	0.500	1.000	0.500	1.000	$\frac{1}{2}(X_1+X_2)$	0.500	1.000		
3	0.449	0.743	0.362	0.920	$\frac{1}{2}(X_1+X_3)$	0.362	0.920	0.449	0.743
4	0.298	0.838	0.298	0.838	$\frac{1}{2}(X_2+X_3)$	0.298	0.838	0.298	0.838
5	0.287	0.697	0.261	0.767	$\frac{1}{2}(X_2+X_4)$	0.231	0.867	0.227	0.881
6	0.215	0.776	0.236	0.706	$\frac{1}{2}(X_2+X_5)$	0.193	0.865	0.184	0.906
7	0.210	0.679	0.218	0.654	$\frac{1}{2}(X_2+X_6)$	0.168	0.849	0.155	0.922
8	0.168	0.743	0.205	0.610	$\frac{1}{2}(X_3+X_6)$	0.149	0.837	0.134	0.934
9	0.166	0.669	0.194	0.572	$\frac{1}{2}(X_3+X_7)$	0.132	0.843	0.118	0.942
10	0.138	0.723	0.186	0.539	$\frac{1}{2}(X_3+X_8)$	0.119	0.840	0.105	0.949
11	0.137	0.663	0.178	0.510	$\frac{1}{2}(X_3+X_9)$	0.109	0.832	0.0952	0.955
12	0.118	0.709	0.172	0.484	$\frac{1}{2}(X_4+X_9)$	0.100	0.831	0.0869	0.959
13	0.117	0.659	0.167	0.461	$\frac{1}{2}(X_4+X_{10})$	0.0924	0.833	0.0799	0.963
14	0.102	0.699	0.162	0.440	$\frac{1}{2}(X_4+X_{11})$	0.0860	0.830	0.0739	0.966
15	0.102	0.656	0.158	0.422	$\frac{1}{2}(X_4+X_{12})$	0.0808	0.825	0.0688	0.969
16	0.0904	0.692	0.154	0.392	$\frac{1}{2}(X_5+X_{12})$	0.0756	0.827	0.0644	0.971
17	0.0901	0.653	0.151	0.389	$\frac{1}{2}(X_5+X_{13})$	0.0711	0.827	0.0605	0.973
18	0.0810	0.686	0.148	0.375	$\frac{1}{2}(X_5+X_{14})$	0.0673	0.825	0.0570	0.975
19	0.0808	0.651	0.145	0.362	$\frac{1}{2}(X_6+X_{14})$	0.0640	0.823	0.0539	0.976
20	0.0734	0.681	0.143	0.350	$\frac{1}{2}(X_6+X_{15})$	0.0607	0.824	0.0511	0.978
∞	$1.57/N$	0.637		0.000	$\frac{1}{2}(P_{25}+P_{15})$	$1.24/N$	0.808		1.000

Tables E.5 and E.6 are reproduced with permission of the McGraw-Hill Companies from Wilfrid J. Dixon and Frank J. Massey Jr., *Introduction to Statistical Analysis*, 3rd Edn, McGraw-Hill Book Company, 1969, p. 488.

Appendix F
Los Angeles Rainfall 1877–1997

Table F.1. Los Angeles Times 4 July 1997

Date	Inches	Date	Inches	Date	Inches
1877–78	21.26	1905–06	18.65	1933–34	14.55
1878–79	11.35	1906–07	19.30	1934–35	21.66
1879–80	20.34	1907–08	11.72	1935–36	12.07
1880–81	13.13	1908–09	19.18	1936–37	22.41
1881–82	10.40	1909–10	12.63	1937–38	23.43
1882–83	12.11	1910–11	16.18	1938–39	13.07
1883–84	38.18	1911–12	11.60	1939–40	19.21
1884–85	9.21	1912–13	13.42	1940–41	32.76
1885–86	22.31	1913–14	23.65	1941–42	11.18
1886–87	14.05	1914–15	17.05	1942–43	18.17
1887–88	13.87	1915–16	19.92	1943–44	19.22
1888–89	19.28	1916–17	15.26	1944–45	11.59
1889–90	34.84	1917–18	13.86	1945–46	11.65
1890–91	13.36	1918–19	8.58	1946–47	12.66
1891–92	11.85	1919–20	12.52	1947–48	7.22
1892–93	26.28	1920–21	13.65	1948–49	7.99
1893–94	6.73	1921–22	19.66	1949–50	10.60
1894–95	16.11	1922–23	9.59	1950–51	8.21
1895–96	8.51	1923–24	6.67	1951–52	26.21
1896–97	16.86	1924–25	7.94	1952–53	9.46
1897–98	7.06	1925–26	17.56	1953–54	11.99
1898–99	5.59	1926–27	17.76	1954–55	11.94
1899–1900	7.91	1927–28	9.77	1955–56	16.00
1900–01	16.29	1928–29	12.66	1956–57	9.54
1901–02	10.60	1929–30	11.52	1957–58	21.13
1902–03	19.32	1930–31	12.53	1958–59	5.58
1903–04	8.72	1931–32	16.95	1959–60	8.18
1904–05	19.52	1932–33	11.88	1960–61	4.85

Table F.1. *continued*

Date	Inches	Date	Inches	Date	Inches
1961–62	18.79	1973–74	14.92	1985–86	17.86
1962–63	8.38	1974–75	14.35	1986–87	7.66
1963–64	7.93	1975–76	7.22	1987–88	12.48
1964–65	13.69	1976–77	12.31	1988–89	8.08
1965–66	20.44	1977–78	33.44	1989–90	7.35
1966–67	22.00	1978–79	19.67	1990–91	11.99
1967–68	16.58	1979–80	26.98	1991–92	21.00
1968–69	27.47	1980–81	8.98	1992–93	27.36
1969–70	7.77	1981–82	10.71	1993–94	8.14
1970–71	12.32	1982–83	31.25	1994–95	24.35
1971–72	7.17	1983–84	10.43	1995–96	12.46
1972–73	21.26	1984–85	12.82	1996–97	12.40

Average 120 Years–14.98 inches

Appendix G
Software Considerations

I. DATA REDUCTION

SAS (Statistical Analysis System, Cory NC) was used exclusively in the development of the examples in this book. However, SAS is not the only software package that yields successful results.

Estimates of Weibull distribution parameters can be made by maximizing the likelihood function:

$$L(x_1, x_2, \cdots x_n; y, \beta, \theta) = \left(\frac{\beta}{\theta^\beta}\right)^n \prod_i (x_i - \gamma)^{\beta-1} \exp\left\{-\sum_i \left(\frac{x_i - \gamma}{\theta}\right)^\beta\right\}$$

(G.1)

Estimates Eq. (1.14) of γ, β, and θ then are called maximum likelihood estimators, or MLE. Any non linear programming procedure (see optimization section below) can be used to find the MLEs for a Weibull distribution.

Source code for a FORTRAN program to calculate MLEs is given by Cohen and Whitten [G.1]. They also discuss alternative techniques for parameter estimation, such as moment estimators [G.2] and Wycoff, Bain, Engelhardt, and Zanakis estimators.

Simple non linear regression analysis can return a set of parameters based on a least-squares analysis, but hypothesis testing should be done to provide a figure of merit to determine whether a good fit has been achieved and whether the data set could have been randomly drawn from a Weibull distribution.

Probability of failure calculations can be done in any programming language that offers a random number generator based on a uniform distribution. The algorithms of Rubinstein [B.1] provide the techniques for generating random variates from a wide variety of distributions.

II. OPTIMIZATION (NON-LINEAR PROGRAMMING)

Today, many procedures for optimizing non-linear criterion functions with constraints are available. A review of these is given by Moré and Wright [C.1].

Many of these procedures are available as UNIX freeware or shareware. The National Institute of Science and Technology (NIST) has a Guide to Available Mathematical Software (GAMS) which can be accessed on the Internet:

http://gams.nist.gov

Optimization problems can also be solved using the computers of Argonne National Laboratory and Northwestern University. Problems can be submitted either from a web page or by e-mail using their templates see:

http://www.mcs.anl.gov/home/otc for details and source code for some routines.

In the past, the authors have used IMSL, (Visual Numerics, Inc. Houston, TX), a large collection of statistical and mathematical subroutines for FORTRAN or C programming languages.

REFERENCES

G.1. Cohen AC, Whitten BJ. Parameter Estimation in Reliability and Life Span Models, Marcel Dekker Inc, 1988, pp. 341–367.
G.2. Cohen AC, Whitten BJ. Parameter Estimation in Reliability and Life Span Models, Marcel Dekker Inc, 1988, pp. 31–46.

Author Index

Abernethy, RB, 27
Agrawal, GK, 182
Aoki, M, 182
Angus, RW, 218

Bain, LJ, 182, 218
Bazovsky, I, 218
Benjamin, JA, 218
Boller, CHR, 134
Bompas-Smith, JH, 218
Bowker, AH, 27

Calabro, SR, 218
Castleberry, G, 134
Cohen, AC, 270
Craver, JS, 27

D'Agostina, RB, 27
Den Hartog, JP, 182
Deutschman, AD, 134
Dieter, GE, 27, 134
DiRoccaferrera, GMF, 182
Dixon, JR, 27, 134, 182
Dixon, WJ, 27
Duffin, RJ, 182

Faires, VM, 134, 182
Faupel, JH, 134
Forrest, P, 134
Fox, RL, 182

Frost, NE, 134
Fry, TR, 134
Furman, TT, 182

Good, IS, 134
Gottfried, BS, 182
Griffel, W, 182
Grover, HJ, 134
Grube, KR, 27

Hagendorf, HC, 134
Hahn, GJ, 218
Hald, A, 27
Harrington, RI, 218
Haugen, EB, 27
Hine, CR, 134
Hodge, JL, 134
Hogg, RV, 27
Horowitz, J, 134

Ireson, WG, 218

Johnson, LG, 27
Johnson, NL, 134
Johnson, RC, 134
Juvinall, RC, 27

King, JR, 27, 218
Kececioglu, DR, 135, 218
Kemeny, JG, 135

Subject Index